Springer Series in Advanced Microelectronics

Volume 58

The Springer Series in Advanced Microelectronics provides systematic information on all the topics relevant for the design, processing, and manufacturing of microelectronic devices. The books, each prepared by leading researchers or engineers in their fields, cover the basic and advanced aspects of topics such as wafer processing, materials, device design, device technologies, circuit design, VLSI implementation, and subsystem technology. The series forms a bridge between physics and engineering and the volumes will appeal to practicing engineers as well as research scientists.

More information about this series at http://www.springer.com/series/4076

Rino Micheloni
Editor

Solid-State-Drives (SSDs) Modeling

Simulation Tools & Strategies

Springer

Editor
Rino Micheloni
Performance Storage Business Unit
Microsemi Corporation
Vimercate
Italy

ISSN 1437-0387 ISSN 2197-6643 (electronic)
Springer Series in Advanced Microelectronics
ISBN 978-3-319-51734-6 ISBN 978-3-319-51735-3 (eBook)
DOI 10.1007/978-3-319-51735-3

Library of Congress Control Number: 2017932535

Printed on acid-free paper

This Springer imprint is published by Springer Nature
The registered company is Springer International Publishing AG
The registered company address is: Gewerbestrasse 11, 6330 Cham, Switzerland

Preface

Solid-State-Drives (SSDs) gained a lot of popularity in the recent few years; compared to traditional HDDs, SSDs exhibit higher speed and reduced power, thus satisfying the tough needs of mobile applications (like smartphones, tablets, and ultrabooks).

A Solid-State-Drive (Chap. 1) is made of a Flash controller plus a bunch of NAND Flash memories (Chap. 2), which are organized in groups called "Flash channels"; each group can have 8, 16 or even more die. Four or eight Flash channels are common in consumer applications but, especially in the enterprise field, the most recent developments span to 32, because of the very high bandwidth requirements. In total, we are easily talking about managing 128, 256 or 512 NAND Flash die. A schematic block diagram of a Solid-State-Drive is sketched in Fig. 1.

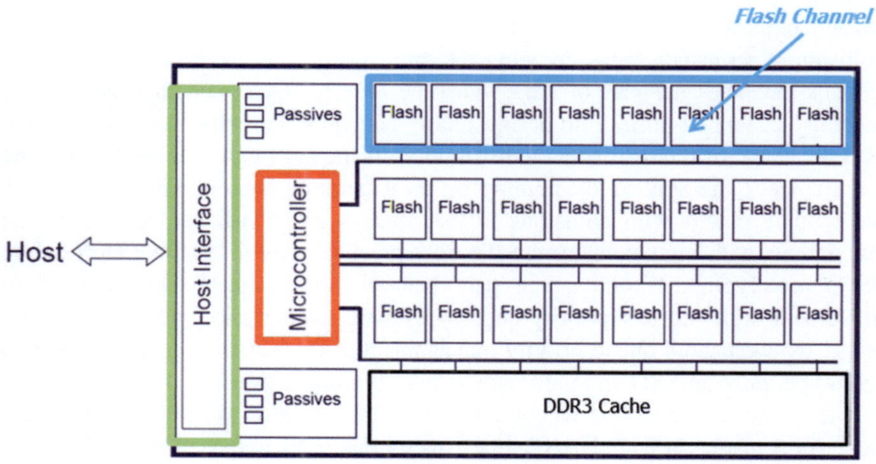

Fig. 1 Solid-State-Drive block diagram

In all SSDs, a Flash controller sits between one or multiple Hosts (i.e. CPUs) and NAND Flash memories, and on each side there are a lot of challenges that designers need to overcome. Moreover, a single controller can have multiple cores, with all the complexity associate with developing a multi-threaded firmware. This book is about how to make simulations of such a complex system, by providing insights on available tools and simulation strategies. As usual, speed and precision don't go hand in hand and it is important to understand when to simulate what, and with which tool. Of course, being able to simulate SSD's performances is mandatory to meet time-to-market, together with product cost and quality.

In order to understand where the challenges are, let's have a closer look at SSD's architectures, by starting from the Host interface. The storage industry has been dominated by *Hard-Disk-Drives* (HDDs) for many decades. HDDs are a rotational media and their data throughput is constrained by the electro-mechanical limitations of the spinning plate and of the servo controllers: this is why HDD's interface speed has remained constant for many years, mainly at 3Gb/s (SATA-II). On the contrary, SSDs don't have mechanical moving parts and they can operate many Flash devices in parallel, thus allowing much higher data transfer rate. As a result, new interfaces have gained momentum in the market: SATA-III and SAS run at 6Gb/s, while PCIe Gen3 is already at 8Gb/s/lane, and PCIe Gen4 is on its way. Of course, there is a practical limit due to area and power, but designers can use parallelism and high speed interfaces to increase performances, and this is why SSD's architectures with 16 or 32 Flash channels are not an exception anymore. When dealing with systems made of hundreds of memory die, queuing effect are not trivial and they strongly depend on the workload and on the scheduling of the data transfers coming from the Host. Transactions could be either sequential or random, and there could be a mixture of read and write operations (typical use cases are 25% read and 75% write, or vice versa). Especially with random traffic, when more requests hit the same NAND die at the same time, a request collision might occur; in other words, the second request in the queue needs to wait until the first one is serviced. This collision effect usually ends up being the root cause for an increased read latency.

Read latencies are part of the *Quality Of Service* (QoS) and they need to be carefully evaluated during the design phase. Not only the worst case matters, but also the shape of the read latency distribution. If the QoS figure doesn't match the requirements, the application running on the Operating System will stall or even hang up. Problem with collisions is that queues are non-linear in nature, and simulations with high statistics are the only practical way to estimate QoS before SSDs prototyping. Is one queue per channel good enough? Do we need to implement high and low priority queues? Can we handle priority at the firmware level? Do we need a queue priority management at the hardware level? Without a prototype, only an SSD simulator can answer these questions.

Let's now have a look at what the Flash controller needs to handle on the Flash side. As a matter of fact, NAND Flash memories have outpaced DRAMs in the technology race. Today we are talking about 256 Gb NAND in a monolithic silicon die, i.e. in less than 150 mm^2. This unbelievable density is not coming for free. In fact, NAND memories are a storage media characterized by a large number errors even at early life and they do require a lot of management algorithms. Besides the well-known Wear Leveling and Garbage Collection algorithms, NANDs in ultra-scaled geometries (below 20 nm) require a plethora of signal processing techniques, from Data Randomization to Read Retry, from V_{TH}-shift to Soft Decoding.

Most of these signal processing techniques translate into read oversampling: more reads means higher probability of collisions, bringing us back into the above mentioned QoS problem. Again, because of the non-linearity, simulations are the only way to predict the impact of these management algorithms on performances (QoS, bandwidth, IOPS).

Error Correction Code (ECC) implementation is another task that SSD's controllers need to perform. BCH codes have been used for many years but the higher and higher NAND BER (Bit Error Rate) has driven towards the adoption of LDPC: these codes are known to be the codes that can get closer to the Shannon limit in real implementations. One drawback of LDPC codes is their need of Soft Information, which has to be extracted from the NAND and then fed into the LDPC decoder by the Flash controller (Chap. 4).

Another important topic is power (Chap. 7). In order to avoid changing existing infrastructures, SSDs need to fit in the same physical space of HDDs and, therefore, in their power envelope. Unfortunately, NAND power consumption is strongly dependent upon the Flash operation (i.e. read, write, erase). For many generations, SSD's power has been over- or under-estimated by adopting the worst case power consumption scenario (i.e. assuming write operations all the time); this approach is running out of steam because it always leads to exceeding power specs with the most recent NAND technologies. In order to correctly evaluate power, it is necessary to estimate the number of operations per time interval and their specific type (read, write, erase). Simulations are again the only practical way to address this problem: when the prototype is ready (especially if the controller is a multi-million dollar ASIC) there are no chances to significantly reduce power, it's too late!

Last but not least, NAND Flash memories wear out, i.e. their performances change over time and they have a limited lifetime. This is even more challenging. Think about the following situation. Let's assume that we have built a prototype of a particular SSD and that performances are within the spec limits. How can we make sure that specs are met when NANDs reach their end of life? Without the proper simulator, the only viable solution is to age a bunch of drives: every time we want to change/test the configuration of a single parameter, we need to consume a number of SSDs, and this is an extremely expensive and time-consuming validation flow.

What's the fastest way to assess QoS? What's the most precise? What's the fastest way to simulate Flash wear-out? … these are just some of the questions that designers need to answer during the development of a modern SSD.

The authors have taken a careful look at both literature and available simulation tools, including popular solutions like VSSIM, NANDFlashSim, and DiskSim. All these solutions are benchmarked against performances of real SSDs, an OCZ VERTEX 120GB and a large enterprise storage platform, that have been measured under different traffic workloads. PROs and CONs of each simulator are analyzed and it is clearly indicated which kind of answers each of them can give and at a what price (Chap. 3).

Over the last few years the authors of this book developed an advanced simulator named "SSDExplorer" (Chap. 3) which has been used for evaluating multiple phenomena, from QoS to Read Retry, from LDPC Soft Information to power, from Flash aging to FTL (Chap. 4). Basically, SSDExplorer is a fine-grained SSD virtual platform that was developed with the following goals in mind:

- offer a RAD tool (Rapid Application Development) for SSD design space exploration;
- accurately predict performance figures of the target architecture;
- offer a wear-out aware framework for complex correction algorithm exploration;
- avoid the overdesign of the resources of the SSD for a target performance.

In Chap. 6, SSDExplorer is used to evaluate the possible impact of emerging non-volatile memories, such as Resistive RAM (RRAM, Chap. 5), on future SSD architectures. Does it make sense to fully replace NANDs with one of the emerging memories? What's the benefit? Is it better to develop a hybrid drive, where a RRAM is used as cache? Most of the new memories are still in the development phase, well before a real mass production; as a matter of fact, simulations are the only way to figure out where a new non-volatile technology can really help, in terms of both performances and cost saving.

In the last chapter of this book, SSD simulators are addressed in a much broader context, i.e. the analysis of what happens when SSDs are connected to the OS (*Operating System*) and to the user application (for example a database search). QoS is again a good example: if a read request is not serviced within a specific time window, the application hangs up, with a very clear impact on the user experience, as the reader can easily imagine. In order to allow this high level simulation, with the required level of precision, SSD simulators need to be directly connected to the end application; if requirements are not met, SSD's firmware can be changed and its impact immediately evaluated. In Chap. 8 authors walk the reader through the full simulation flow of a real system-level by combining SSDExplorer with the QEMU virtual platform. The reader will be impressed by the level of know-how and the combination of models that such simulations are asking for.

In this short introduction we have shown several reasons and examples of why developing a Solid-State-Drive requires a solid simulation strategy and a set of reliable simulation tools. SSDs are very complex, they have to manage several NAND Flash memories in parallel, without forgetting the more and more stringent requirements of end user applications. This book provides an overview of the state-of-the-art simulation techniques, together with an outlook of the challenges that designers will have to address in the coming few years.

Vimercate, Italy Rino Micheloni

Contents

Editor and Contributors

About the Editor

Dr. Rino Micheloni is Fellow at Microsemi Corporation where he currently runs the Non-Volatile Memory Lab in Milan, with special focus on NAND Flash. Prior to joining Microsemi, he was Fellow at PMC-Sierra, working on NAND Flash characterization, LDPC, and NAND Signal Processing as part of the team developing Flash controllers for PCIe SSDs. Before that, he was with IDT (Integrated Device Technology) as Lead Flash Technologist, driving the architecture and design of the BCH engine in the world's 1st PCIe NVMe SSD controller. Early in his career, he led Flash design teams at ST Microelectronics, Hynix, and Infineon; during this time, he developed the industry's first MLC NOR device with embedded ECC technology and the industry's first MLC NAND with embedded BCH.

Rino is IEEE Senior Member, he has co-authored more than 50 publications, and he holds 278 patents worldwide (including 124 US patents). He received the STMicroelectronics Exceptional Patent Award in 2003 and 2004, and the Qimonda IP Award in 2007.

Rino has published the following books with Springer: *3D Flash Memories* (2016), *Inside Solid State Drives* (2013), *Inside NAND Flash Memories* (2010), *Error Correction Codes for Non-Volatile Memories* (2008), *Memories in Wireless Systems* (2008), and *VLSI-Design of Non-Volatile Memories* (2005).

Contributors

Luca Crippa Microsemi Corporation, Vimercate, Italy

Alessia Marelli Microsemi Corporation, Vimercate, Italy

Rino Micheloni Performance Storage Business Unit, Microsemi Corporation, Vimercate, Italy

Piero Olivo Dipartimento di Ingegneria, Università degli Studi di Ferrara, Ferrara, Italy

Cristian Zambelli Dipartimento di Ingegneria, Università degli Studi di Ferrara, Ferrara, Italy

Lorenzo Zuolo Dipartimento di Ingegneria, Università degli Studi di Ferrara, Ferrara, Italy

Chapter 1
Solid State Drives (SSDs)

Rino Micheloni and Luca Crippa

Abstract *Solid-state drives* (SSDs) are unanimously considered the enabling factor for bringing enterprise storage performances to the next level. Indeed, the rotating-storage technology of *Hard Disk Drives* (HDDs) can't achieve the access-time required by applications where response time is the critical factor. On the contrary, SSDs are based on solid state memories, namely NAND Flash memories: in this case, there aren't any mechanical parts and random access to stored data can be much faster, thus addressing the above mentioned needs. In many applications though, the interface between host processors and drives remains the performance bottleneck. This is why SSD's interface has evolved from legacy storage interfaces, such as SAS and SATA, to PCIe, which enables a direct connection of the SSD to the host processor. In this chapter we give an overview of the SSD's architecture by describing the basic building blocks, such as the Flash controller, the Flash File System (FFS), and the most popular I/O interfaces (SAS, SATA and PCIe).

1.1 Introduction

Solid State Drives (SSDs) promise to greatly enhance enterprise storage performance. While electromechanical *Hard Disk Drives* (HDDs) have continuously ramped in capacity, the rotating-storage technology doesn't provide the access-time or transfer-rate performance required in demanding enterprise applications, including on-line transaction processing, data mining, and cloud computing. Client applications are also in need of an alternative to electromechanical disk drives that can deliver faster response times, use less power, and fit into smaller mobile form factors.

Flash-memory-based SSDs can offer much faster random access to data and faster transfer rates. Moreover, SSD's capacity is now at the point where solid state

R. Micheloni (✉) · L. Crippa
Performance Storage Business Unit, Microsemi Corporation, Vimercate, Italy
e-mail: rino.micheloni@ieee.org

L. Crippa
e-mail: luca.crippa@ieee.org

R. Micheloni (ed.), *Solid-State-Drives (SSDs) Modeling*,
Springer Series in Advanced Microelectronics 58,
DOI 10.1007/978-3-319-51735-3_1

drives can serve as rotating-disk replacements. But in many applications the interface between host and drives remains the performance bottleneck. SSDs with legacy storage interfaces, such as SAS and SATA, are proving useful, and PCI-Express (PCIe) SSDs will further increase performance and improve responsiveness, being directly connected to the host processor.

1.2 SSD's Architecture

A basic block diagram of a solid state drive is shown in Fig. 1.1. In addition to memories and a Flash controller, there are usually other components. For instance, an external DC-DC converter can be added in order to drive the internal power supply, or a quartz can be used for a better clock precision. Of course, reasonable filter capacitors are inserted for stabilizing the power supply. It is also very common to have an array of temperature sensors for power management reasons. For data caching, a fast DDR memory is frequently used: during a write access, the cache

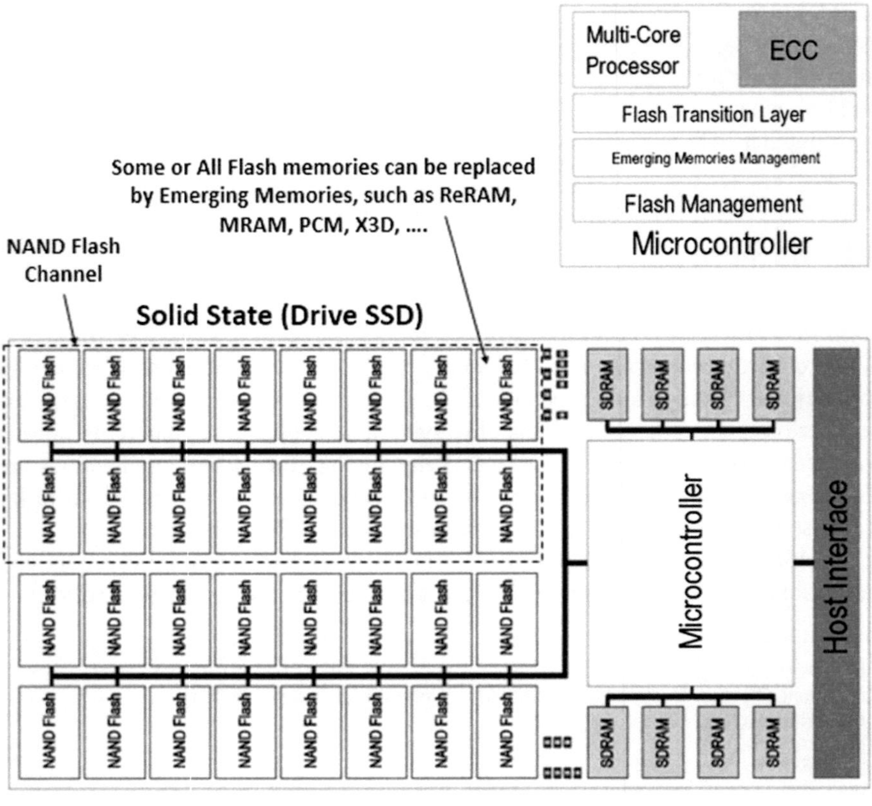

Fig. 1.1 Block diagram of an SSD

is used for storing data before their transfer to the Flash. The benefit is that data updating, e.g. of routing tables, is faster and does not wear out the Flash.

A typical memory system is composed of several NAND memories [1]. Typically, an 8-bit bus [2, 3], usually called "channel", is used to connect different memories to the controller (Fig. 1.1). It is important to underline that multiple Flash memories in a system are both a means for increasing storage density and read/write performances [4].

Operations on a channel can be interleaved, which means that a second chip can be addressed while the first one is still busy. For instance, a sequence of multiple write operations can be directed to a channel, addressing different NANDs, as shown in Fig. 1.2: in this way, the channel utilization is maximized by pipelining the data load phase. In fact, while the program operation takes place inside a memory chip, the corresponding Flash channel is free. The total number of Flash channels is a function of the target application, but tens of channels are becoming quite common. Thanks to interleaving, given the same Flash programming time, SSD's throughput greatly improves.

The memory controller is responsible for scheduling the accesses to the memory channels. The controller uses dedicated engines for the low level communication protocol with the Flash.

Moreover, it is clear that the data load phase is not negligible compared to the program operation (the same comment is valid for data output): therefore, increasing I/O interface speed is another smart way to improve performances: DDR-like interfaces are discussed in more details in Chap. 2. As the speed increases, more NAND can be operated in parallel before saturating the channel. For instance, assuming a target of 30 MB/s, 2 NANDs are needed with a minimum DDR frequency of about 50 MHz. Given a page program time of 200 μs, at 50 MHz four NANDs can operate in interleaved mode, doubling the write throughput. Of course, power consumption is another metric to be carefully considered.

After this high level overview of the SSD's architecture, let's move to the heart of the architecture: the memory (Flash) controller.

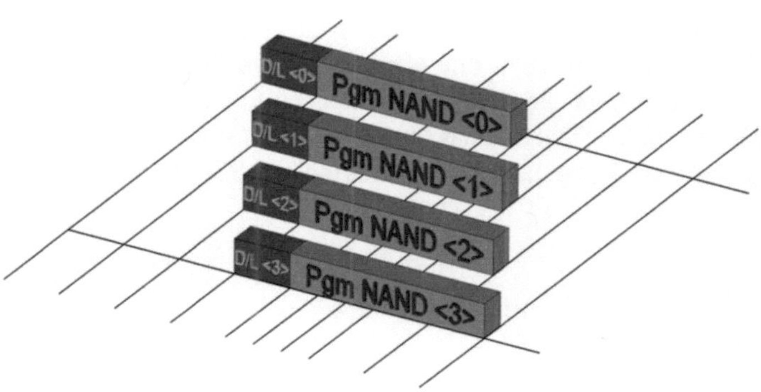

Fig. 1.2 Interleaved operations on one Flash channel

1.3 Flash Controller

A memory controller has two fundamental tasks:

1. provide the most suitable interface and protocol towards both the host and the Flash memories;
2. efficiently handle data, maximizing transfer speed, data integrity and retention of the stored information.

In order to carry out such tasks, an application specific device is designed, embedding a standard processor—usually 8–16 bits—together with dedicated hardware to handle time-critical tasks.

Generally speaking, the memory controller can be divided into four parts, which are implemented either in hardware or in firmware (Fig. 1.3).

Moving from the host to the Flash, the first part is the host interface, which implements the required industry-standard protocol (PCIe, SAS, SATA, etc.), thus ensuring both logical and electrical interoperability between SSDs and hosts. This block is a mix of hardware (buffers, drivers, etc.) and firmware, which decodes the command sequence invoked by the host and handles the data flow to/from the Flash memories.

The second part is the *Flash File System* (FFS) [5]: that is, the file system which enables the use of SSDs like magnetic disks. For instance, sequential memory access on a multitude of sub-sectors which constitute a file is organized by linked lists (stored on the SSD itself), which are used by the host to build the *File Allocation Table* (FAT). The FFS is usually implemented in form of firmware inside the controller, each sub-layer performing a specific function. The main functions are: *Wear leveling Management*, *Garbage Collection* and *Bad Block Management*. For all these functions, tables are widely used in order to map sectors and pages from the logical domain to the physical domain (Flash Translation Layer or FTL) [6, 7], as shown in Fig. 1.4. The top row is the logical view of the memory, while the bottom row is the physical one. From the host perspective, data are transparently written and overwritten inside a given logical sector: due to Flash limitations, overwrite on the same page is not possible; therefore, a new page (sector) must be allocated in the physical block, and the previous one is marked as invalid. It is clear that, at some point in time, the current physical block becomes full and a second one (coming from the pool of "buffer" blocks) has to take over that logic address.

The required translation tables are stored on the SSD itself, thus reducing the overall storage capacity.

1.4 Wear Leveling

Usually, not all the data stored within the same memory location change with the same frequency: some data are often updated while others don't change for a very

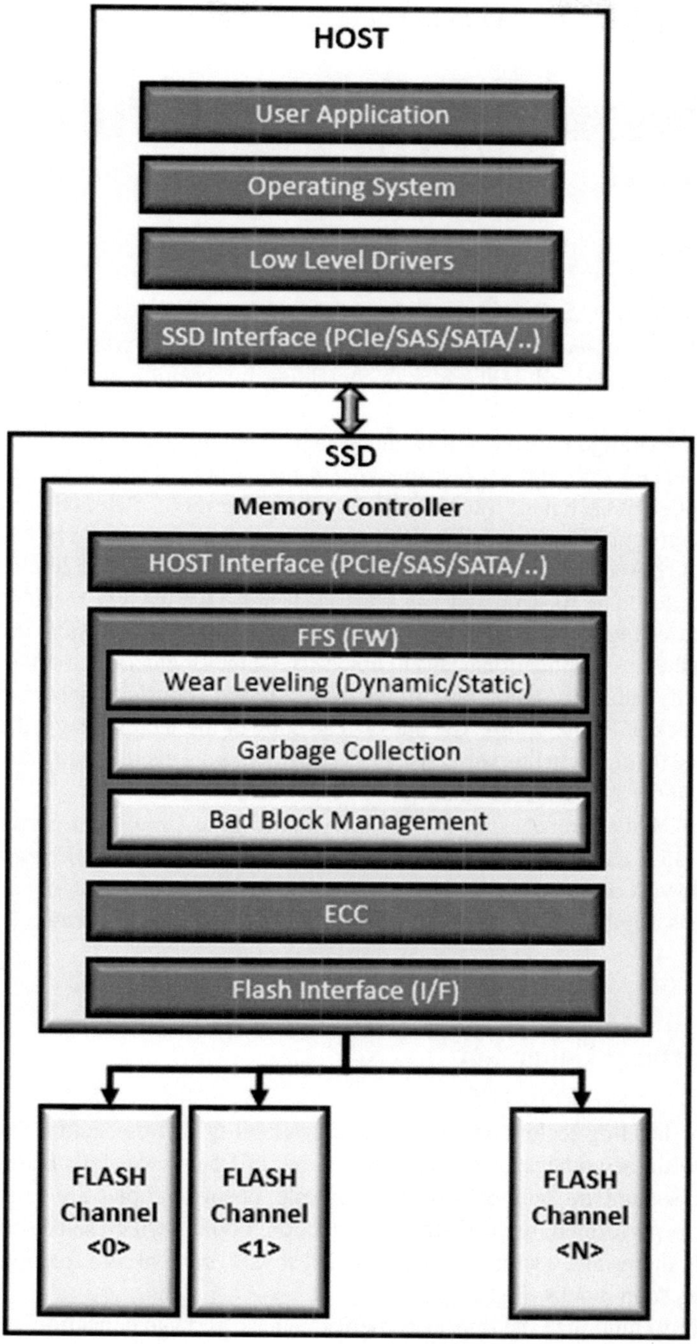

Fig. 1.3 High level view of a Flash controller

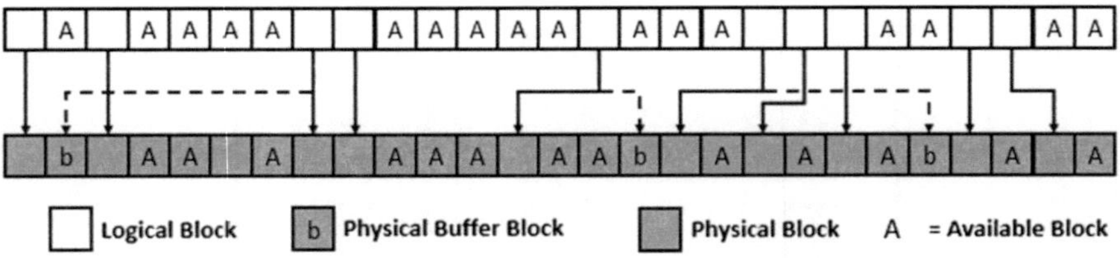

Fig. 1.4 Logical to physical block management

long time—in the extreme case, for the whole life of the device. It's clear that the blocks containing frequently-updated information are stressed with a larger number of write/erase cycles, while the blocks containing information updated very rarely are much less stressed.

In order to mitigate disturbs, it is important to keep the aging of each page/block as minimum and as uniform as possible: that is, the number of both read and program cycles applied to each page must be monitored. Furthermore, the maximum number of allowed program/erase cycles for a block (i.e. its endurance) should be considered: in case SLC NAND memories are used, this number is in the order of 20–30 k cycles, which is reduced to 10–15 k and 1–5 k for MLC and TLC NAND, respectively.

Wear Leveling techniques exploit the concept of logical to physical translation: each time the host application needs to update the same (logical) sector, the memory controller dynamically maps that sector to a different (physical) sector, of course keeping track of the mapping. The out-of-date copy of the sector is tagged as invalid and eligible for erase. In this way, all the physical blocks are evenly used, thus keeping the aging under a reasonable value.

Two kinds of approaches are possible: Dynamic Wear Leveling is normally used to follow up a user's request of update, writing to the first available erased block with the lowest erase count; with Static Wear Leveling every block, even the least modified, is eligible for re-mapping as soon as its aging deviates from the average value.

1.5 Garbage Collection

Both wear leveling techniques rely on the availability of free sectors that can be filled up with the updates: as soon as the number of free sectors falls below a given threshold, sectors are "compacted" and multiple, obsolete copies are deleted. This operation is performed by the Garbage Collection module, which selects the blocks containing the invalid sectors, it copies the latest valid content into free sectors, and then erases such blocks (Fig. 1.5).

In order to minimize the impact on performances, garbage collection can be performed in background. The aging uniformity driven by wear leveling distributes wear out stress over the whole array rather than on single hot spots. Hence, given a specific

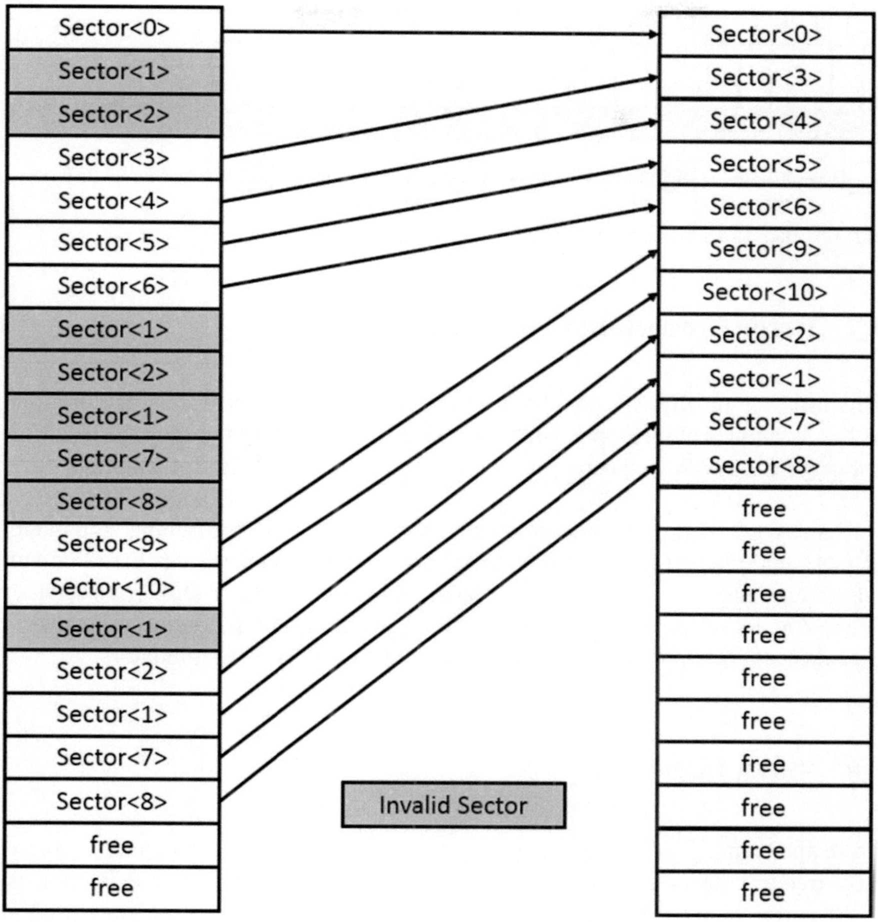

Fig. 1.5 Garbage collection

workload and usage time, the bigger the memory density, the lower the wear out per cell.

1.6 Bad Block Management

No matter how smart the Wear Leveling algorithm is, an intrinsic limitation of NAND Flash memories is represented by the presence of the so-called *Bad Blocks* (BB), i.e. blocks which contain one or more locations whose reliability is not guaranteed. The *Bad Block Management* (BBM) module creates and maintains a map of bad blocks, as shown in Fig. 1.6: this map is created in the factory and then updated during SSD's lifetime, whenever a block becomes bad.

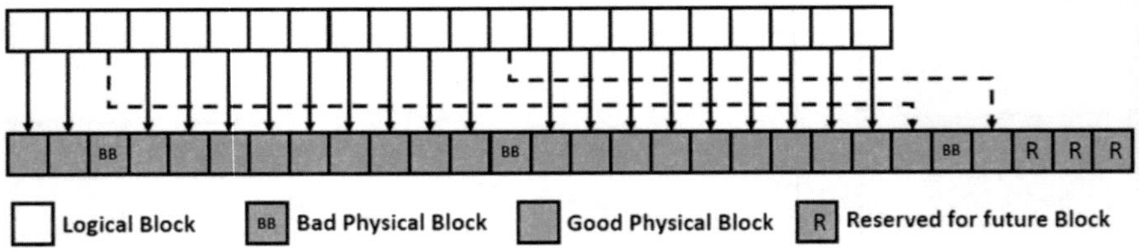

Fig. 1.6 Bad Block Management (BBM)

1.7 Error Correction Code (ECC)

This task is typically executed by a hardware accelerator inside the memory controller. Examples of memories with embedded ECC were also reported [8–10]. Most popular ECC codes, correcting more than one error, are Reed-Solomon and BCH [11].

NAND raw BER gets worse generation after generation approaching, as a matter of fact, the Shannon limit. As a consequence, correction techniques based on soft information processing are becoming more and more popular: LDPC (*Low Density Parity Check*) codes are an example of this soft information approach. Chapter 4 provides details about how these codes can be handled by SSD simulators.

1.8 SSD's Interfaces

There are 3 main interface protocols used to connect SSDs into server and/or storage infrastructure: *Serial Attached SCSI* (SAS), *Serial ATA* (SATA) and PCI-Express. PCI-Express based SSDs deliver the highest performances and are mainly used in server based deployments as a plug-in card inside the server itself. SAS SSDs deliver pretty good level of performances and are used in both high-end servers and mid-range and high-end storage enclosures. SATA based SSDs are used mainly in client applications and in entry-level and mid-range server and storage enclosures.

1.9 SAS and SATA

Serial Attached SCSI (SAS) is a communication protocol traditionally used to move data between storage devices and host. SAS is based on a serial point-to-point physical connection. It uses a standard SCSI command set to drive device communications. Today, SAS based devices most commonly run at 6 Gbps, but 12 Gbps SAS are available too. On the other side, SAS interface can also be run at slower speeds—1.5 Gbps and/or 3 Gbps to support legacy systems.

SAS offers backwards-compatibility with second-generation SATA drives. The T10 technical committee of the *International Committee for Information Technology Standards* (INCITS) develops and maintains the SAS protocol; the *SCSI Trade Association* (SCSITA) promotes the technology.

Serial ATA (SATA or *Serial Advanced Technology Attachment*) is another interface protocol used for connecting host bus adapters to mass storage devices, such as hard disk drives and solid state drives. Serial ATA was designed to replace the older parallel ATA/IDE protocol. SATA is also based on a point-to-point connection. It uses ATA and ATAPI command sets to drive device communications. Today, SATA based devices run either at 3 or 6 Gbps.

Serial ATA industry compatibility specifications originate from the Serial ATA International Organization [12] (aka. SATA-IO).

A typical SAS eco-system consists of SAS SSDs plugged into a SAS backplane or a host bus adapter via a point to point connection, which, in turn, is connected to the host microprocessor either via either an expander or directly, as shown in Fig. 1.7.

Each expander can support up to 255 connections to enable a total of 65535 (64k) SAS connections. Indeed, SAS based deployments enable the usage of a large number of SAS SSDs in a shared storage environment.

SAS SSDs are built with two ports. This dual port functionality allows host systems to have redundant connections to SAS SSDs. In case one of the connections to the SSD is either broken or not properly working, host systems still have the second port that can be used to maintain continuous access to the SAS SSD. In enterprise applications where high availability is an absolute requirement, this feature is key.

SAS SSDs also support hot-plug: this feature enables SAS SSDs to be dynamically removed or inserted while the system is running; it also allows automatic detection of

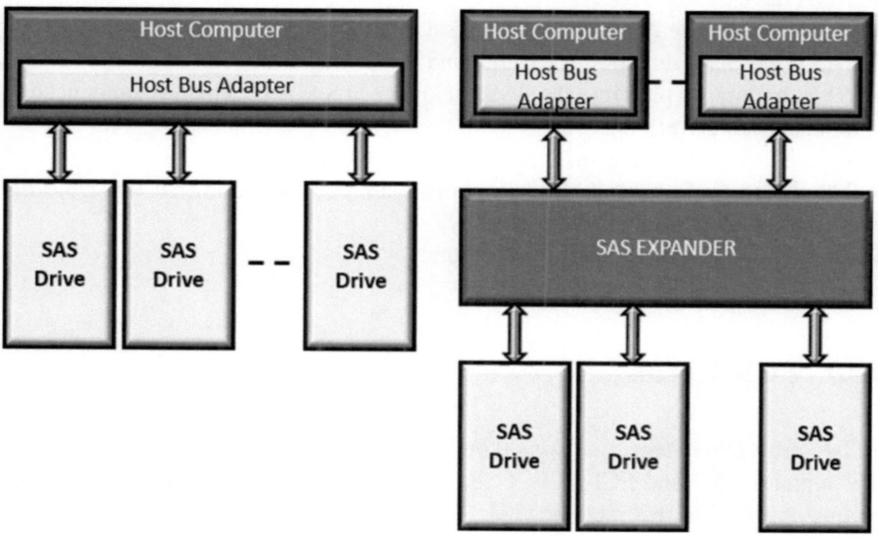

Fig. 1.7 SAS connectivity

newly inserted SAS SSDs. In fact, while a server or storage system is running, newly inserted SAS SSDs can be dynamically configured and put in use. Additionally, if SAS SSDs are pulled out of a running system, all the in-flight data that were already committed to the host system are stored inside the SAS drive, and can be accessed at a later point, when the SSD is powered back on.

Differently from SAS, a typical SATA infrastructure consists of SATA SSDs point-to-point connected to a host bus adapter driven by the host microprocessor. In addition, SATA drives are built with a single port, unlike SAS SSDs. These two main differences make SATA based SSDs a good fit for entry or mid-range deployments and consumer applications.

The SATA protocol supports hot-plug; however, not all SATA drives are designed for it. In fact, hot-plug requires specific hardware (i.e. additional cost) to guarantee that committed data, which are actually still in-flight, are safely stored during a power drop.

It is worth highlighting that SATA drives may be connected to SAS backplanes, but SAS drives can't be connected to SATA backplanes. Of course, this another reason for the broad SATA penetration in the market.

Similarities between SAS and SATA are:

- Both types plug into the SAS backplane;
- The drives are interchangeable within a SAS drive bay module;
- Both are long proven technologies, with worldwide acceptance;
- Both employ point-to-point architecture;
- Both are hot pluggable.

Differences between SAS and SATA are:

- SATA devices are cheaper;
- SATA devices use the ATA command set, SAS the SCSI command set;
- SAS drives have dual port capability and lower latencies;
- While both types plug into the SAS backplane, a SATA backplane cannot accommodate SAS drives;
- SAS drives are tested against much more rigid specifications;
- SAS drives are faster and offer additional features, like variable sector size, LED indicators, dual port, and data integrity;
- SAS supports link aggregation (wide port).

1.10 PCI-Express

PCI-Express (*Peripheral Component Interconnect Express*) or PCIe is a bus standard that replaced PCI and PCI-X. PCI-SIG (PCI *Special Interest Group*) creates and maintains the PCIe specification [13].

PCIe is used in all computer applications including enterprise servers, consumer personal computers (PC), communication systems, and industrial applications.

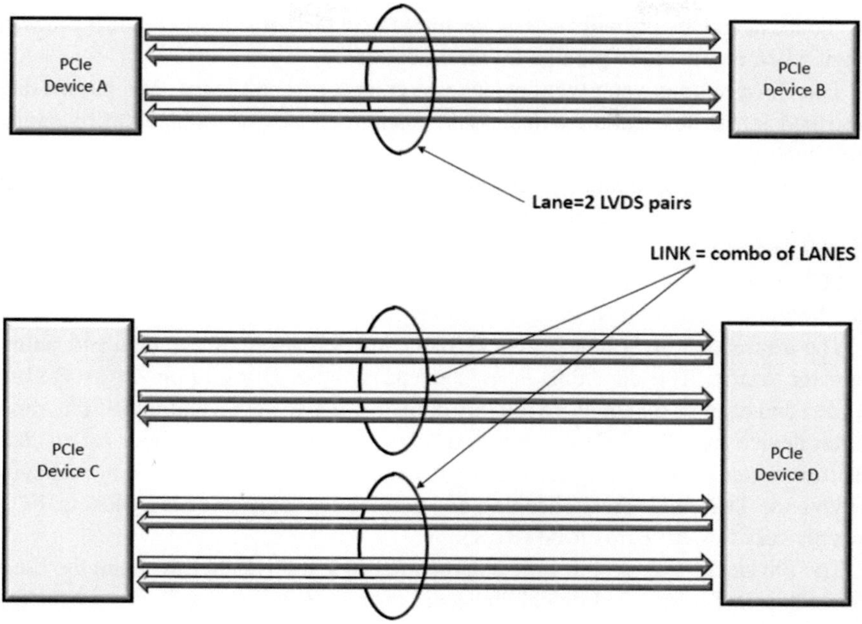

Fig. 1.8 PCI Express lane and link. In Gen2, 1 lane runs at 5Gbps/direction; a 2-lane link runs at 10 Gbps/direction

Unlike the older PCI bus topology, which uses shared parallel bus architecture, PCIe is based on point-to-point topology, with separate serial links connecting every device to the root complex (host). Additionally, a PCIe link supports full-duplex communication between two endpoints. Data can flow upstream (UP) and downstream (DP) simultaneously. Each pair of these dedicated unidirectional serial point-to-point connections is called a *lane*, as depicted in Fig. 1.8. The PCIe standard is constantly under improvement, with PCIe 3.0 being the latest version of the standard available in the market (Table 1.1). The standardization body is currently working on defining Gen4.

Other important features of PCIe include power management, hot-swappable devices, and the ability to handle peer-to-peer data transfers (sending data between two end points without routing through the host) [14]. Additionally, PCIe simpli-

Table 1.1 Throughput of different PCIe generations

PCIe version	Year introduced	Throughput per lane
PCIe 1.0 (Gen1)	2003	250 MB/s
PCIe 2.0 (Gen2)	2007	500 MB/s
PCIe 3.0 (Gen3)	2010	1 GB/s

fies board design by utilizing a serial technology, which drastically reduces the wire count when compared to parallel bus architectures.

The PCIe link between two devices can consist of 1–32 lanes. The packet data is striped across lanes, and the lane count is automatically negotiated during device initialization.

The PCIe standard defines slots and connectors for multiple widths: $\times 1$, $\times 4$, $\times 8$, $\times 16$, $\times 32$. This allows PCIe to serve lower throughput, cost-sensitive applications as well as performance-critical applications.

PCIe uses a packet-based layered protocol, consisting of a transaction layer, a data link layer, and a physical layer, as shown in Fig. 1.9.

The transaction layer handles packetizing and de-packetizing of data and status-message traffic. The data link layer sequences these *Transaction Layer Packets* (TLPs) and ensures that they are reliably delivered between two endpoints. If a transmitter device sends a TLP to a remote receiver device and a CRC error is detected, the transmitter device gets a notification back. The transmitter device automatically replays the TLP. With error checking and automatic replay of failed packets, PCIe ensures very low *Bit Error Rate* (BER).

The Physical Layer is split in two parts: the Logical Physical Layer and the Electrical Physical Layer. The Logical Physical Layer contains logic gates for processing packets before transmission on the Link, and processing packets from the Link to the Data Link Layer. The Electrical Physical Layer is the analog interface of the Physical Layer: it consists of differential drivers and receivers for each lane.

TLP assembly is shown in Fig. 1.10. Header and Data Payload are TLP's core information: Transaction Layer assembles this section based on the data received from the application software layer. An optional End-to-End CRC (ECRC) field can be appended to the packet. ECRC is used by the ultimate targeted device of this packet to check for CRC errors inside Header and Data Payload. At this point, the Data Link Layer appends a sequence ID and local CRC (LCRC) field in order to protect the ID. The resultant TLP is forwarded to the Physical Layer which concatenates a Start and End framing characters of 1 byte each to the packet. Finally, the packet is encoded and differentially transmitted on the Link by using the available Lanes.

Today, PCIe is a high volume commodity interconnect used in virtually all computers, from consumer laptops to enterprise servers, as the primary motherboard technology that interconnects the host CPU with on-board ICs and add-on peripheral expansion cards.

1.11 The Need for High Speed Interfaces

Processor vendors have continued to ramp the performance of individual processor cores, to combine multiple cores in one chip, and to develop technologies that can closely couple multiple chips in multi-processor systems. Ultimately, all of the cores in such a scenario need access to the same storage subsystem.

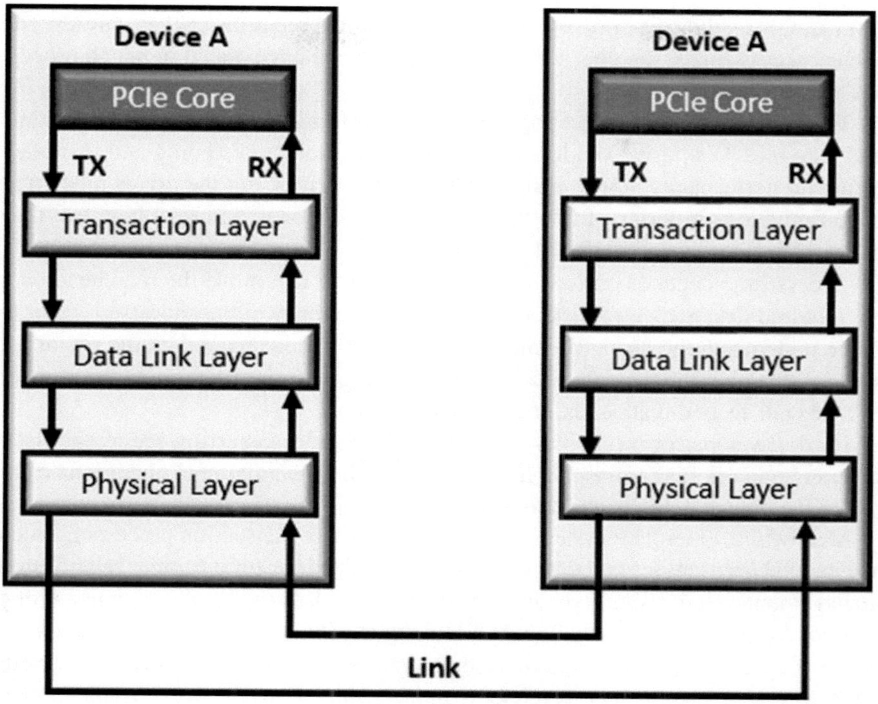

Fig. 1.9 PCIe Layered architecture

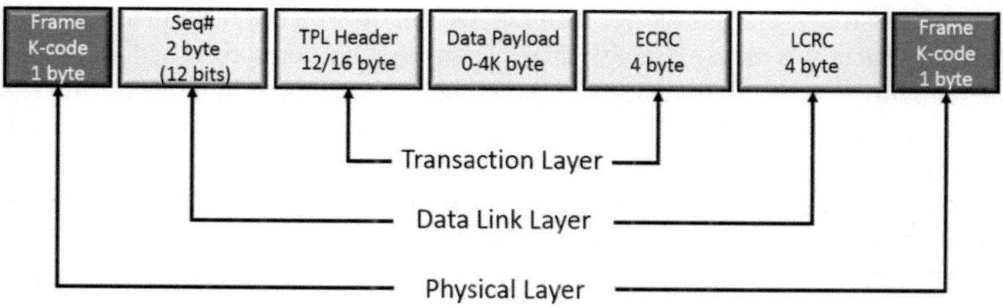

Fig. 1.10 Transaction Layer Packet (TLP) assembly

Enterprise IT managers are eager to utilize the multiprocessor systems because they have the potential of boosting the number of I/O operations per second (IOPS) that a system can process and also the number of IOPS per Watt. This multiprocessing computing capability offers better IOPS relative to cost and power consumption—assuming the processing elements can get access to the data in a timely fashion. Active processors waiting on data waste time and money.

There are, of course, multiple levels of storage technology in a system that ultimately feed code and data to each processor core. Generally, each core includes local cache memory that operates at core speed. Multiple cores in a chip share a second-level and, sometimes, a third-level cache. And DRAM feeds the caches. DRAM and caches access-times, together with data-transfer speed have scaled to match processor's performance.

The issue is the performance gap between DRAM and HDD in terms of access time and data rate. Disk/drive vendors have done a great job at designing and manufacturing higher-capacity, lower-cost-per-Gbyte disks/drives; but the drives inherently have limitations in terms of how fast they can access data, and then how fast they can transfer these data to DRAM.

Access time depends on how quickly a hard drive can move the read head over the required data track on a disk, and the rotational latency of the addressed sector to move underneath the head. The maximum transfer rate is dictated by the rotational speed of the disk and the data encoding scheme: together they determine the number of bytes per second that can be read from the disk.

Hard drives perform relatively well in reading and transferring sequential data. But random seek operations add latency. And even sequential read operations can't match the data appetite of the latest processors.

Meanwhile, enterprise systems that perform on-line transaction processing, such as financial transactions and data mining (e.g. applications for customer relationship management) require highly random access to data. Also cloud computing has strong random requirements, especially when looking at virtualization, which expands the scope of different applications that a single system has active at any one time. Every microsecond of latency directly relates to money, utilization of processors and system power.

Fortunately, Flash memories can help reducing the performance gap between DRAM and HDD. Flash is slower than DRAM but offers a lower cost per Gbyte of storage. That cost is more expensive than disk storage, but enterprises will gladly pay the premium because Flash also offers much better throughput in terms of Mbyte/s and faster access to random data, resulting in better cost-per-IOPS compared to rotating storage.

Neither the legacy disk-drive form factor nor the interface is ideal for Flash-based storage. SSD manufacturers can pack enough Flash devices in a 2.5-in form factor to easily exceed the power profile developed for disk drives. And Flash can support higher data transfer rates than even the latest generation of disk interfaces.

Let's examine the disk interfaces more closely (Fig. 1.11). The third-generation SATA and SAS support 600 Mbyte/s throughput, and drives based on those interfaces have already found usage in enterprise systems. While those data rates support the

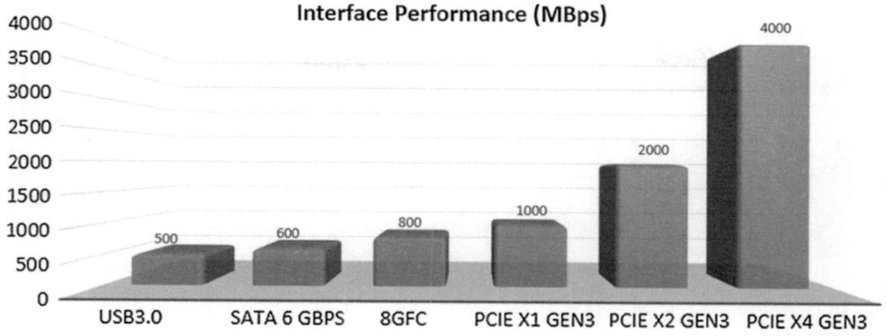

Fig. 1.11 Interface performance. PCIe improves overall system performance by reducing latency and increasing throughput

fastest electromechanical drives, new NAND Flash architectures and multi-die Flash packaging deliver aggregate Flash bandwidth that exceeds the throughput capabilities of SATA and SAS interconnects. In short, the SSD performance bottleneck has shifted from the storage media to the host interface. Therefore, many applications need a faster host interconnect to take full advantage of Flash storage.

The PCIe host interface can overcome this storage performance bottleneck and deliver unparalleled performance by attaching the SSD directly to the PCIe host bus. For example, a 4-lane (x4) PCIe Generation 3 (Gen3) link can deliver 4 GByte/s data rates. Simply put, PCIe meets the desired storage bandwidth. Moreover, the direct PCIe connection can reduce system power and slash the latency that's attributable to the legacy storage infrastructure.

Clearly, an interface such as PCIe can handle the bandwidth of a multi-channel Flash storage subsystem and can offer additional performance advantages. SSDs that use a disk interface also suffer latency added by a storage-controller IC that handles disk I/O. PCIe devices connect directly to the host bus, thus eliminating the architectural layer associated with the legacy storage infrastructure. The compelling performance of PCIe SSDs has resulted in system manufacturers placing PCIe drives in servers as well as in storage arrays to build tiered storage systems that accelerate applications while improving cost-per-IOPS.

The benefits of using PCIe as a storage interconnect are clear. You can achieve over 6x the data throughput compared to SATA or SAS. You can eliminate components such as host bus adapters and SerDes ICs on the SATA and SAS interfaces—saving money and power at the system level. And PCIe moves the storage closer to the host CPU reducing latency, as shown in Fig. 1.12.

Latency, IOPS, bandwidth, power, interface speed, number of channels, NAND-type (SLC, MLC, TLC, QLC) are all parameters that SSD designers need to take into account to meet their target specifications at minimum cost. Looking forward,

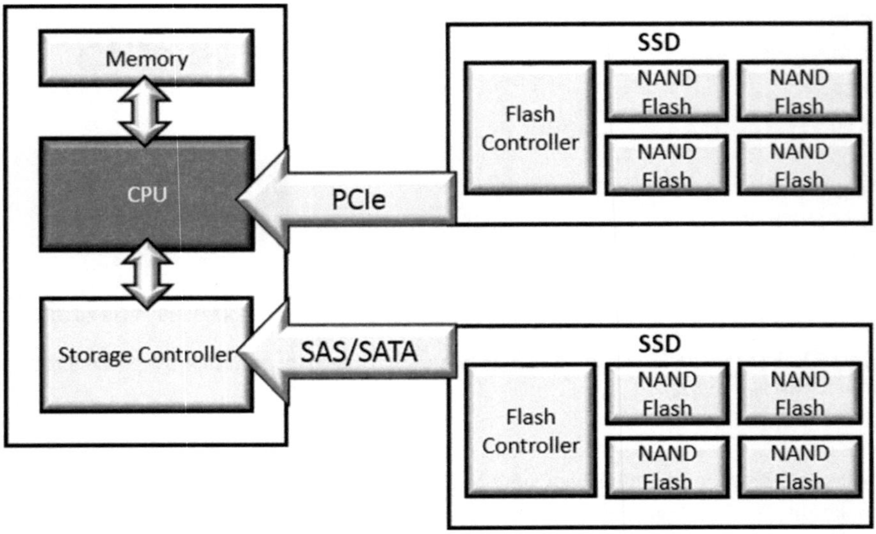

Fig. 1.12 PCIe SSD versus SAS/SATA SSD

emerging memories will be part of the game too. It is clear that, given this number of variables, a simple approach based on hardware prototyping is hard to pursue, especially when looking at the time-to-market. Therefore, SSD simulators become a must have and they will be covered in great details in the following chapters.

References

1. G. Campardo, R. Micheloni, D. Novosel, "VLSI-Design of Non-Volatile Memories", Springer-Verlag, 2005.
2. www.onfi.org
3. www.jedec.org
4. C. Park et al., "A High Performance Controller for NAND Flash-based Solid State Disk (NSSD)", IEEE Non-Volatile Semiconductor Memory Workshop NVSMW, pp. 17–20, Feb. 2006.
5. A. Kawaguchi, S. Nishioka, and H. Motoda. "A flash-memory based file system", Proceedings of the USENIX Winter Technical Conference, pp. 155–164, 1995.
6. J. Kim, J. M. Kim, S. Noh, S. L. Min, and Y. Cho. A space-efficient flash translation layer for compactflash systems. IEEE Transactions on Consumer Electronics, 48(2):366–375, May 2002.
7. S.-W. Lee, D.-J. Park, T.-S. Chung, D.-H. Lee, S.-W. Park, and H.-J. Songe. "FAST: A log-buffer based ftl scheme with fully associative sector translation", 2005 US-Korea Conference on Science, Technology, & Entrepreneurship, August 2005.
8. T. Tanzawa, T. Tanaka, K. Takekuchi, R. Shirota, S. Aritome, H. Watanabe, G. Hemink, K. Shimizu, S. Sato, Y. Takekuchi, and K. Ohuchi, "A compact on-chip ECC for low cost Flash memories," IEEE J.Solid-State Circuits, vol. 32, pp. 662–669, May 1997.

9. G. Campardo, R. Micheloni et al.,"40-mm^2 3-V-only 50-MHz 64-Mb 2-b/cell CHE NOR Flash memory" – IEEE Journal of Solid-State Circuits, Vol. 35, No. 11, Nov. 2000, pp. 1655–1667.
10. R. Micheloni et al., "A 4Gb 2b/cell NAND Flash Memory with Embedded 5b BCH ECC for 36MB/s System Read Throughput", IEEE International Solid-State Circuits Conference Dig. Tech. Papers, pp. 142–143, Feb. 2006.
11. R. Micheloni, A. Marelli, R. Ravasio, "Error Correction Codes for Non-Volatile Memories", Springer-Verlag, 2008.
12. http://www.sataio.org
13. www.pcisig.com
14. R. Budruk, D. Anderson, T. Shanley, "PCI Express System Architecture", Mindshare 2003.

Chapter 2
NAND Flash Memories

Rino Micheloni and Luca Crippa

Abstract NAND Flash memories are the storage media used inside *Solid State Drives* (SSDs). Indeed, a single drive for enterprise applications can contain up to hundreds of Flash chips. Flash memories are non-volatile in the sense that they can retain the information even when powered off, but they wear out, i.e. their performances change over time and they have a limited lifetime. Therefore, taking time to learn the basics of the Flash technology is a necessary step for people who have to deal with SSDs. After an introduction about the matrix array, this chapter walks the reader through the NAND basic functionalities, i.e. program, erase, and read operations. The second part of the chapter is devoted to the digital interface of NAND memories, with a detailed description of the communication protocol, for both synchronous (DDR) and asynchronous (legacy) modes.

2.1 Introduction

Semiconductor memories can be divided into two major categories: RAM, acronym for *Random Access Memories*, and ROM, acronym for *Read Only Memories*: RAM loses its content when power supply is switched off, while ROM virtually holds it forever. A third category is NVM, which stands for *Non-Volatile Memories*, whose content can be electrically altered but it is also preserved when power supply is removed. These memories are more flexible than the original ROM, whose content is defined during manufacturing and cannot be changed by the user in the field.

The history of non-volatile memories began in the Seventies, with EPROMs (*Erasable Programmable Read Only Memories*). In the early 90s, with the introduction of non-volatile Flash memories into portable products like mobile phones,

R. Micheloni (✉)
Performance Storage Business Unit, Microsemi Corporation, Vimercate, Italy
e-mail: rino.micheloni@ieee.org

L. Crippa
Microsemi Corporation, Vimercate, Italy
e-mail: luca.crippa@ieee.org

R. Micheloni (ed.), *Solid-State-Drives (SSDs) Modeling*,
Springer Series in Advanced Microelectronics 58,
DOI 10.1007/978-3-319-51735-3_2

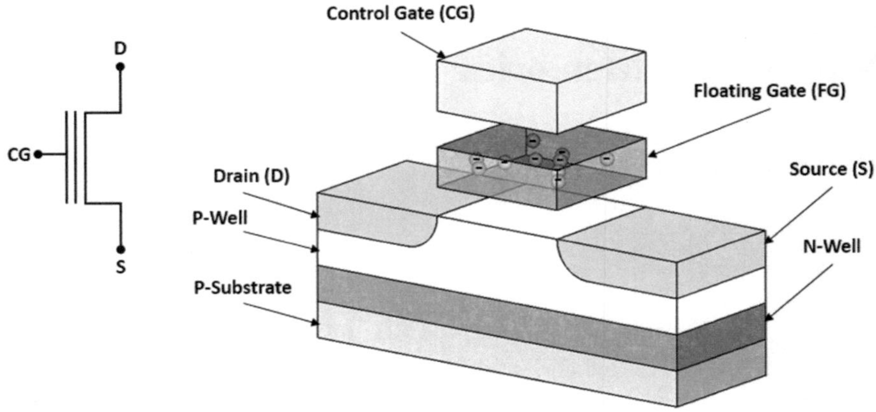

Fig. 2.1 (*right*) Floating gate memory cell and (*left*) its schematic symbol

USB keys, digital cameras and so on, the NVM market has experienced a stunning increase.

The most popular planar Flash memory cell is based on the *Floating Gate* (FG) technology [1], whose cross section is shown in Fig. 2.1. A MOS transistor is built with two overlapping gates: the first one is completely surrounded by oxide and it is known as "floating gate", while the second one is the actual gate terminal. The isolated gate constitutes an excellent "trap" for electrons, thus enabling long data retention. The operations performed to inject and remove electrons from the floating gate are called program and erase, respectively. These operations modify the threshold voltage V_{TH} of the memory cell, which, in most of the cases, is a n-type MOS transistor. By applying fixed voltages to cell's terminals, it is possible to discriminate two storage levels: when the gate voltage is higher than cell's V_{TH}, the cell is ON ("1"), otherwise it is OFF ("0").

It is worth mentioning that, due to issues in scaling floating gate technologies, another kind of NAND is gaining traction, especially with 3D architectures: charge trap. Details about this technology, together with a deep dive of the various 3D options, can be found in [2]. The good news is that all the following considerations about basic operations hold true for both floating gate and charge trap.

2.2 NAND Flash Array

Memory cells are packed together to form a matrix, in order to optimize silicon area occupation. Depending on how the cells are organized inside the matrix, it is possible to distinguish between NAND and NOR [3] Flash memories. When looking at *Solid State Drives* (SSDs), NAND is definitely a synonym for Flash.

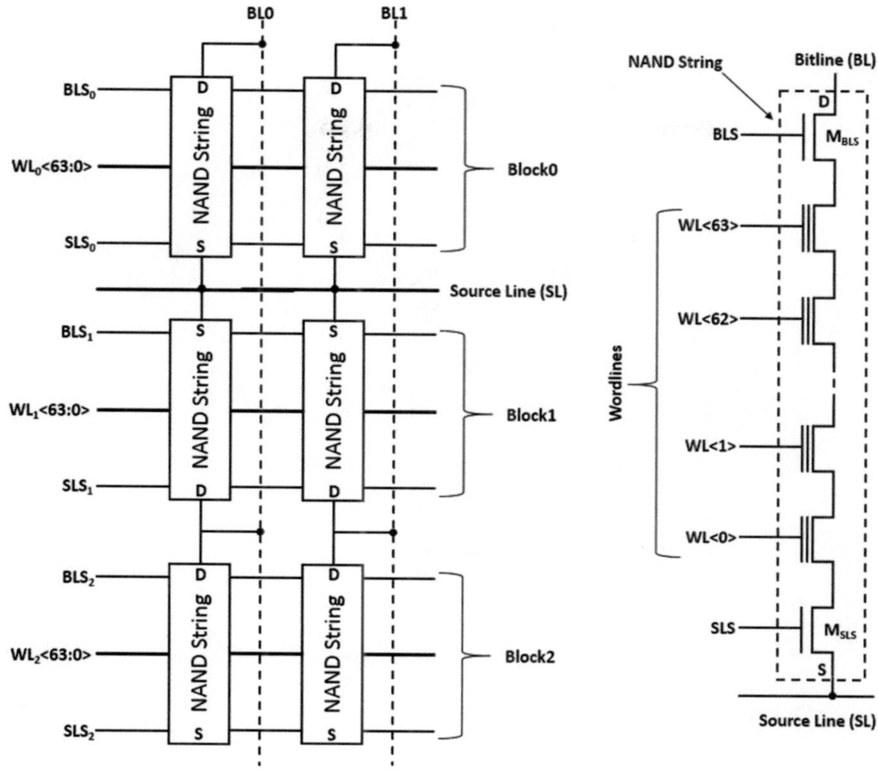

Fig. 2.2 NAND string (*left*) and NAND array (*right*)

In the NAND string, cells are connected in series, in groups of 64, 128 or even more [4], as shown in Fig. 2.2. Two select transistors, M_{SL} and M_{DL}, sit at the edges of the string, to ensure the connections to source line and bitline (BL). Each NAND string shares the bitline contact with another string. Control gates are connected through wordlines (WLs).

Logical pages are made of cells belonging to the same wordline. The number of pages per wordline is related to the storage capabilities of the memory cell. Depending on the number of storage levels, Flash memories are referred to in different ways: SLC memories store 1 bit per cell, MLC memories store 2 bits per cell, TLC memories store 3 bits per cell, and QLC memories store 4 bits per cell.

All the NAND strings sharing the same group of wordlines are erased together, thus forming a so-called Flash Block. In Fig. 2.2 there are 3 blocks: by using a bus representation, one block is made of WL_X <63:0>.

A NAND Flash device is not just the memory array, as sketched in Fig. 2.3. The Row Decoder is located between the planes (i.e. sub-arrays): this circuit has the task of properly biasing all the wordlines belonging to the selected NAND string. All the bitlines are connected to sense amplifiers (Sense Amp), sometimes called

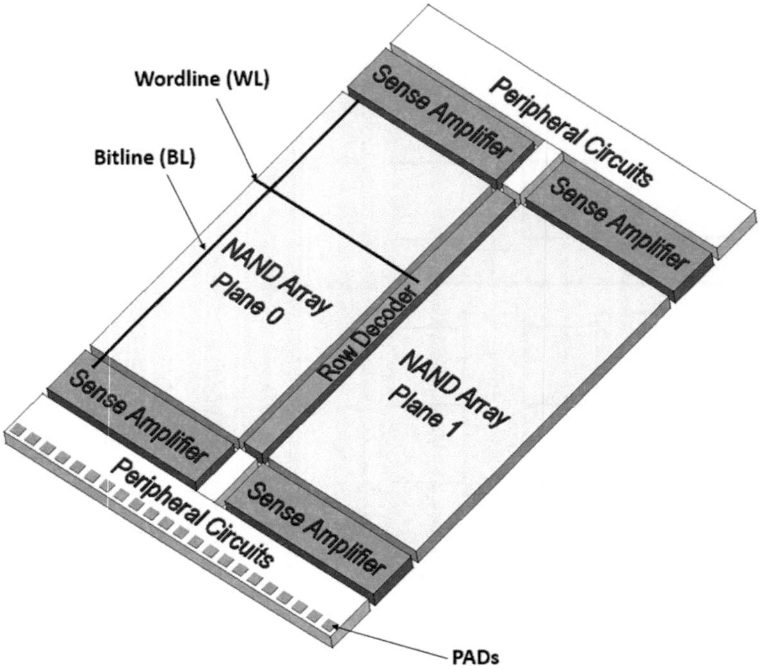

Fig. 2.3 Block diagram of a NAND Flash memory

Page Buffers. The purpose of sense amplifiers is to convert the analog value of the current sunk by the memory cell into a digital bit. In the peripheral area there are charge pumps and voltage regulators for producing voltages higher than the external power supply VDD, logic circuits, and redundancy structures. PADs are used to communicate with the external world.

2.3 Flash Basic Operations

This section briefly describes the basic NAND functionalities: read, program, and erase.

2.3.1 Read

During a read operation (Fig. 2.4), cell's gate is driven at V_{READ} (0 V), while the other cells are biased at $V_{PASS,R}$ (usually 4÷5 V), so that they can act as pass-transistors, regardless the value of their threshold voltages. In fact, an erased Flash cell has a

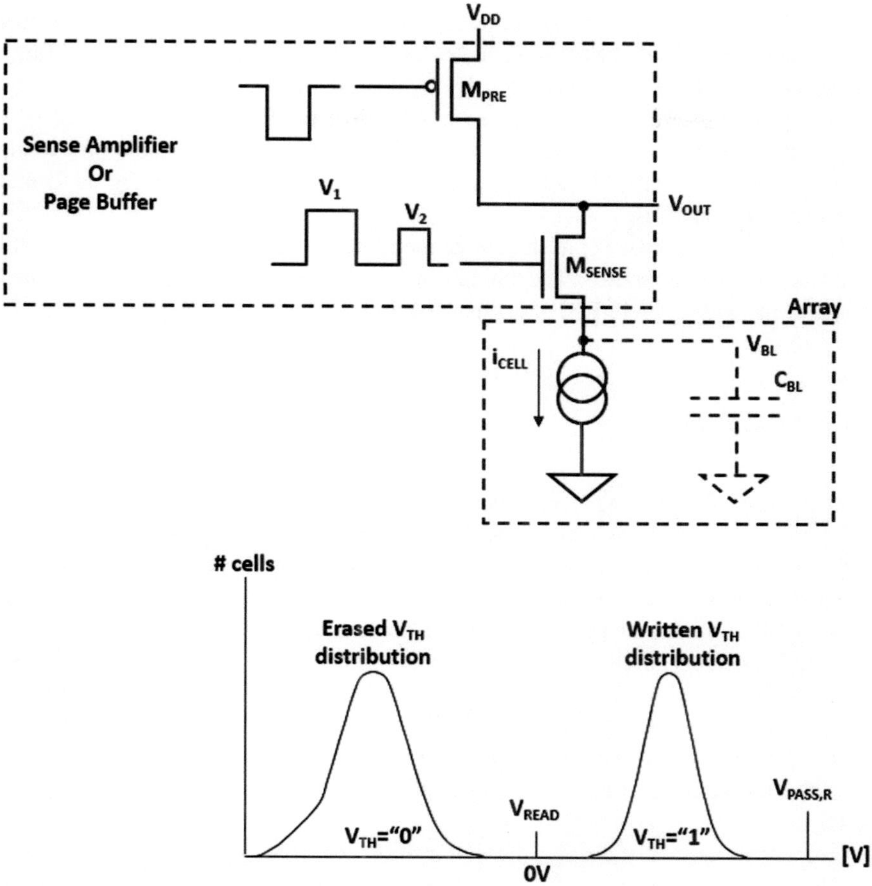

Fig. 2.4 NAND string biasing during read and SLC V_{TH} distributions

V_{TH} smaller than 0 V; vice versa, a written cell has a positive V_{TH} but, however, smaller than 4 V. In practice, by biasing the gate of the selected cell with a voltage equal to 0 V, the series of all the cells will conduct current only if the addressed cell is erased.

The read technique is based on charge integration, exploiting the bitline parasitic capacitor. This capacitor is precharged at a fixed value (usually 1÷1.2 V): only if the cell is erased and sinks current, then the capacitor is discharged. Several circuits exist to detect the bitline parasitic capacitor state: the structure depicted in the inset of Fig. 2.4 is one of the most common. The bitline parasitic capacitor is indicated with C_{BL} while the NAND string is modeled with a current generator.

During the charge of the bitline, the gate of the PMOS transistor M_P is kept grounded, while the gate of the NMOS transistor M_N is forced at a fixed value V_1. Typical value for V_1 is around 2 V. At the end of the transient the bitline has a voltage

$V_{BL} = V_1 - V_{THN}$, where V_{THN} indicates the threshold voltage value of M_N. At this point, transistors M_N and M_P are switched off and C_{BL} is free to discharge. After a time t_{VAL}, the gate of M_N is biased at $V_2 < V_1$, usually 1.6÷1.4 V.

If t_{VAL} is long enough to discharge the bitline voltage V_{BL} under the value ($V_2 - V_{THN}$), then M_N turns on and the voltage of node OUT (V_{OUT}) becomes equal to the one of the bitline. Finally, the analog voltage V_{OUT} is converted into a digital format by using simple latches.

There are alternatives [3] to the above mentioned technique but they are all still based on a charge integration. The main difference is the fact that they use a dedicated capacitor instead of the bitline parasitic capacitor.

2.3.2 *Program*

Programming of NAND memories exploits the quantum-effect of electron tunneling in the presence of a strong electric field (Fowler-Nordheim tunneling [5]).

During programming, the number of electrons crossing the oxide is a function of the electric field: in fact, the stronger the field, the higher the injection probability. Thus, in order to improve program performances, it is essential to have high electric fields and, therefore, high voltages. This requirement is one of the main drawbacks of this program method, since the oxide degradation is accelerated by these voltages.

The main advantage is the extremely low current consumption, in the range of few nA per cell. This is what makes the Fowler-Nordheim mechanism suitable for a parallel programming of many cells as required by the long NAND page size (e.g. 16 kB).

In order to precisely change the cell's threshold voltage, a *Program & Verify* algorithm [3] is adopted: verify is used to check whether the cell has reached the target distribution or not.

To trigger the injection of electrons into the floating gate, the following voltages are applied, as shown in Fig. 2.5:

- V_{DD} on the gate of the drain selector;
- $V_{PASS,P}$ (8 ÷ 10 V) on the unselected gates;
- V_{PGM} (20 ÷ 25 V) on the selected gate (to be programmed);
- GND on the gate of the source selector;
- GND on the bitlines to be programmed;
- V_{DD} on other bitlines.

The so-called *self-boosting* mechanism [3] prevents the cells sitting on the same wordline from undergoing an undesired programming. The basic idea is to exploit the high voltages generated during the programming operation to increase the potential of the region underneath the tunnel oxide (inset of Fig. 2.5). This is accomplished by leveraging the parasitic capacitor of the memory cell itself.

When bitlines are biased at V_{DD}, drain selectors are diode-connected and the corresponding bitlines are left floating. By applying $V_{PASS,P}$ to the unselected wordlines,

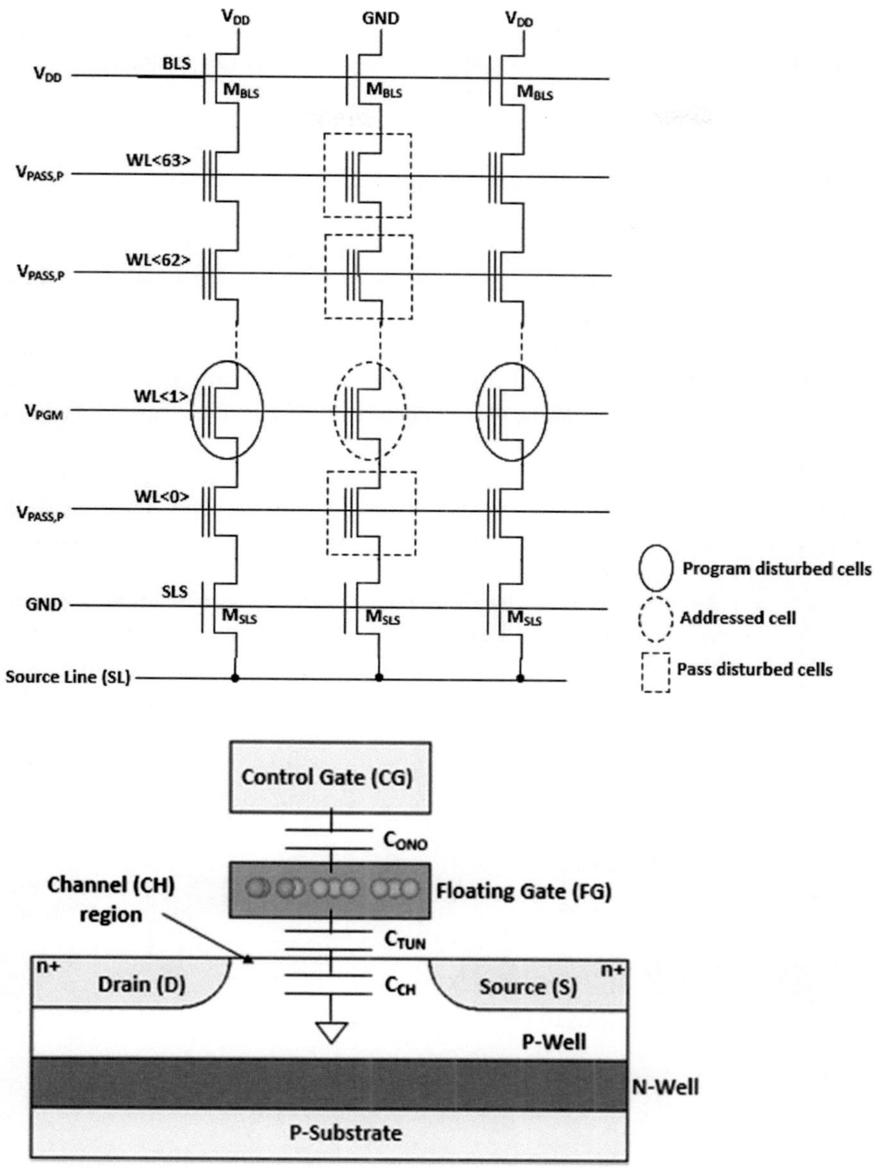

Fig. 2.5 NAND string biasing during programming. The self-boosting mechanism is used to inhibit programming of unselected cells

cells parasitic capacitors boost the potential of the channel, thus reducing the voltage drop across the tunnel oxide and, hence, inhibiting tunneling phenomena.

2.3.3 *Erase*

The NAND array is placed in a triple-well structure, as shown in Fig. 2.6a. Usually, each plane has its own triple-well. The source terminal is shared by all the blocks; in this way the matrix is more compact.

The electrical erase takes place by biasing the iP-well at high voltage and keeping the wordlines of the NAND block to be erased grounded (Fig. 2.6c). Therefore, NAND technologies don't need negative voltages. The physical mechanism is again the Fowler-Nordheim tunneling. As the iP-well is common to all the blocks, erase of unselected blocks is prevented by leaving their wordlines floating. In this way, when the iP-well is charged, the potential of the floating wordlines raises as well, thanks to the capacitive coupling between the control gates and the iP-well (i.e. Fowler-Nordheim tunneling is inhibited).

Figure 2.6b sketches the erase algorithm phases. NAND specifications are quite aggressive in terms of erase time. Therefore, Flash vendors try to erase the block

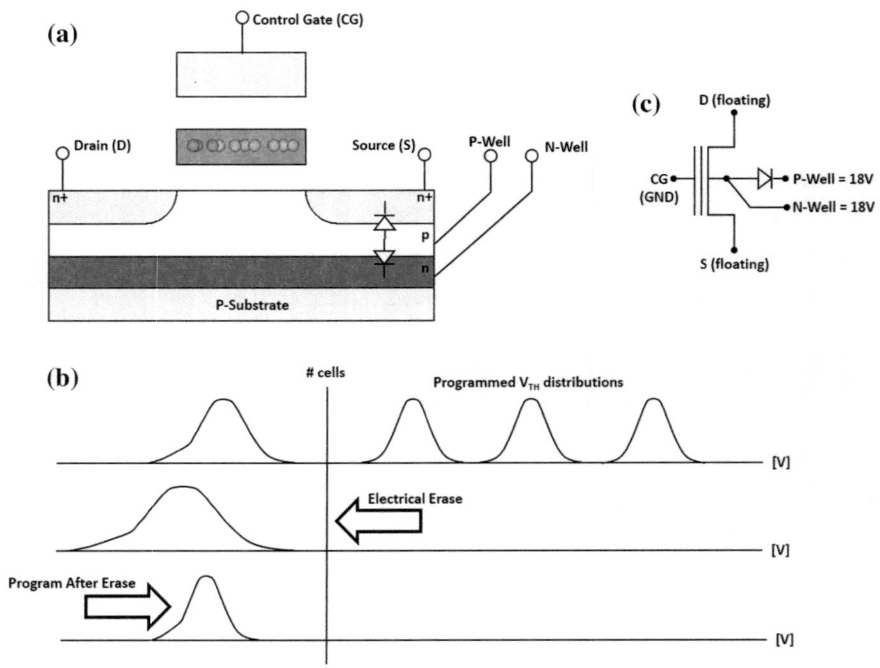

Fig. 2.6 **a** NAND matrix in triple-well; **b** erase algorithm; **c** erase biasing on the selected block

content in few erase steps. As a consequence, a very high electric field is applied to the matrix during the *Electrical Erase* phase. As a matter of fact, erased distribution is deeply shifted towards negative V_{TH} values. In order to minimize floating gate coupling, a *Program After Erase* (PAE) phase is introduced, with the goal of placing the distribution near the 0 V limit (of course, guaranteeing the appropriate read margin).

Technology shrink asks for even more sophisticated erase algorithms, especially for 3–4 bit/cell devices. In fact, reliability margins shrink with the technology node: a more precise, and therefore time consuming, PAE has to be adopted, to contain the erased distribution width. In summary, erase time is becoming longer and longer, generation after generation, reaching values in the range of $10 \div 20$ ms (especially with 3D memories).

2.4 NAND Flash Memory Map

NAND Flash memories are divided in pages and blocks, as sketched in Fig. 2.7. A block is the smallest erasable unit. Generally speaking, there are a power of two blocks within any device, plus few extra blocks to compensate for Bad Blocks. Each block contains multiple pages. A page is the smallest addressable unit for reading and writing. Each page is composed of main area and spare area (Fig. 2.7). Main area can be either 8 or 16 kB. Spare area is used for ECC [6, 7] and system (firmware) logical pointers and it is in the order of a couple of hundreds bytes per 4 kB of main area.

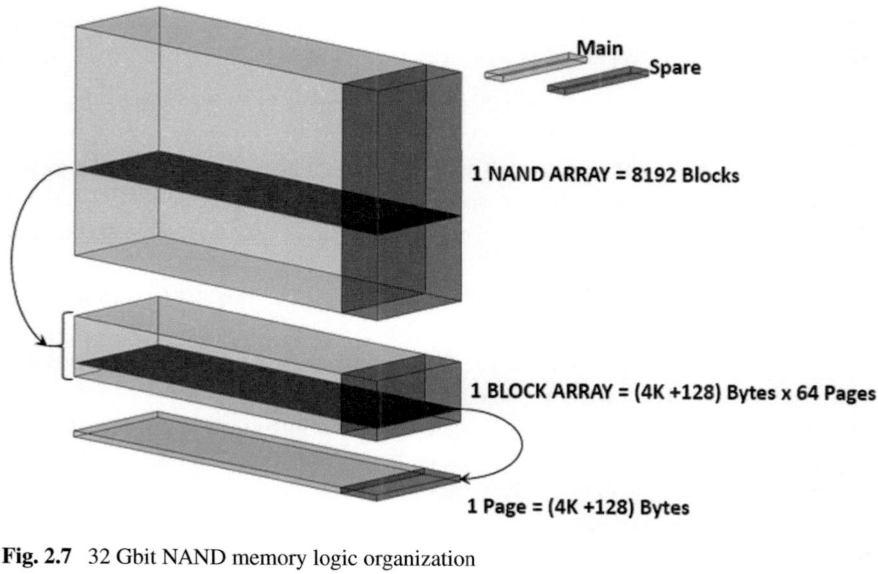

Fig. 2.7 32 Gbit NAND memory logic organization

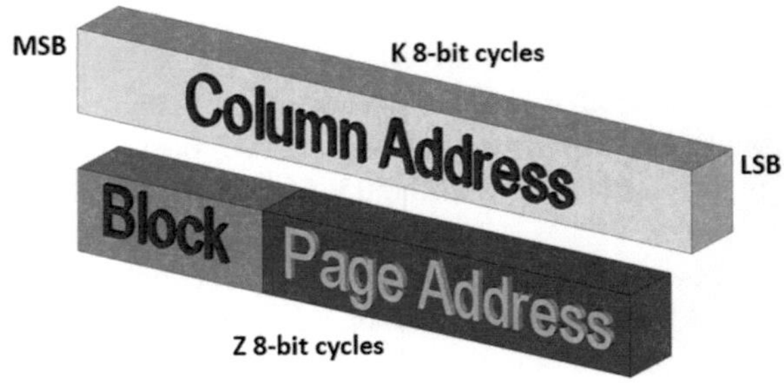

Fig. 2.8 NAND addressing

Every time we want to execute an operation on a NAND device, we must specify the memory address. The address is split in row and column (Fig. 2.8). Row address identifies the addressed page, while column address is used to identify a byte inside the page. When both row and column addresses are required, column address is given first, 8 bits per address cycle. The first cycle contains the least significant bits. Row and column addresses cannot share the same address cycle.

The row address identifies the block and the page involved in the operation. Page address occupies the least significant bits.

2.5 NAND Commands

NAND devices communicate with the external world by means of pins. These pins are monitored by the *Command Interface* (CI) embedded in the NAND device, which has the task to understand what the host wants to execute.

For many years, the asynchronous interface (Fig. 2.9) has been the only available option for NAND devices, and it is described here below.

- CE# : it is the Chip Enable signal. This input signal is "1" when the device is in stand-by mode, otherwise it is always "0".
- R/B# : it is the Ready/Busy signal. This output signal is used to indicate the target status. When low, the target has an operation in progress.
- RE# : it is the Read Enable signal. This input signal is used to enable serial data output.
- CLE : it is the Command Latch Enable. This input is used by the host to indicate that the bus cycle is used to input the command.
- ALE : it is the Address Latch Enable. This input is used by the host to indicate that the bus cycle is used to input the addresses.
- WE# : it is the Write Enable. This input signal controls the latching of input data. Data, command and address are latched on the rising edge of WE#.
- WP# : it is the Write Protect. This input signal is used to disable Flash array program and erase operations.
- DQ<7:0> : these input/output signals represent the data bus.

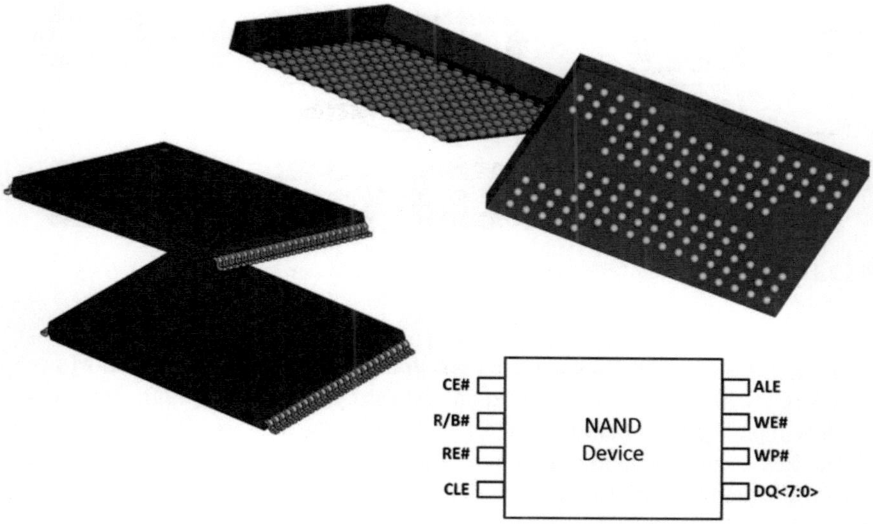

Fig. 2.9 TSOP and BGA packages and related pinout (*right*)

2.5.1 Read Operation

The read function retrieves the data stored at a specified address. In order to accomplish this goal, NAND must recognize when a read request comes in, together with the related addresses. After the "busy time", necessary to perform the internal read algorithm, NAND memories are ready to output data. Based on the device pin signals, the NAND Command Interface is able to understand when a command is issued, when an address is issued, and when it must output data.

Figure 2.10 shows a command cycle ("Cmd"). CI recognizes a "Cmd" cycle if CLE is high. In this case, the 8-bit value on DQs represents the command code.

Figure 2.11 shows address cycles. Generally, all the operations need the addresses where they have to act. The address length depends on the operation and on the

Fig. 2.10 Command cycle ("Cmd"): CLE is high, all other input signals are low

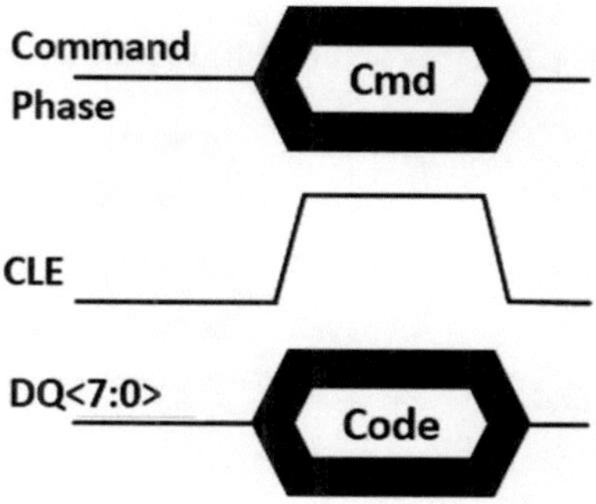

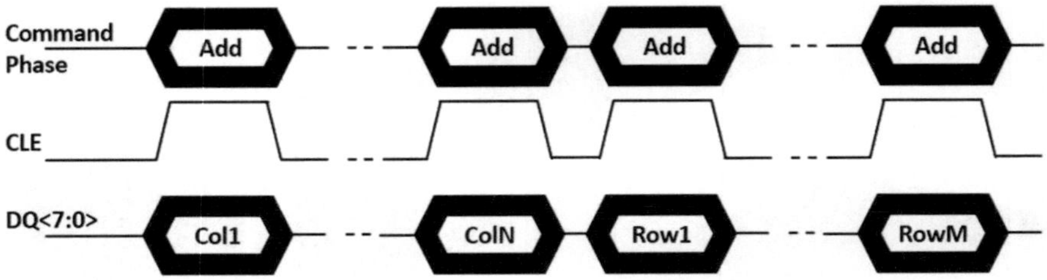

Fig. 2.11 Address cycle: ALE is high, all other inputs signals are low

capacity of the NAND; N cycles are used to input column addresses and M cycles are for row addresses. CI recognized an address cycle if ALE is high. Meanwhile, all other input signals are low and the DQs value is the address.

The last command phase used by the read operation is the data out, shown in Fig. 2.12. Data out is performed by toggling signal RE#: at every cycle a new byte of data is available on DQs.

These basic cycles are used by the NAND to decode and perform every operation. Figure 2.13 shows the full command sequence for a read operation. The first cycle is used to issue the read command "RD" (e.g. 00h). After the command cycle a number of cycles is used to provide the addresses. Column addresses are given first, followed by row addresses. All the pins (ALE, CLE, RE#) not present in the figure must be driven as described above. Code "RDC" (Read Command Confirm, e.g. 30h) is used

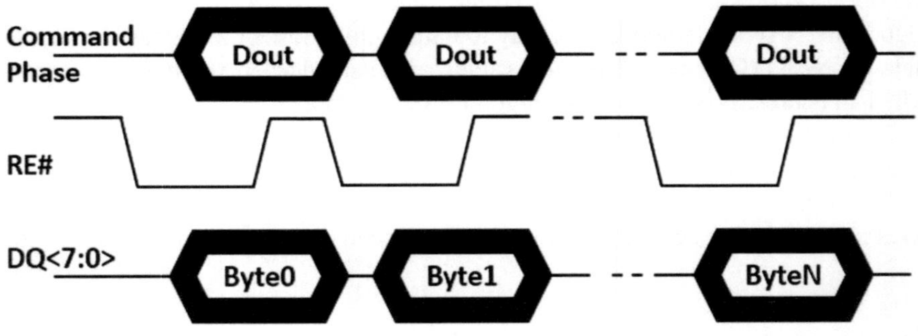

Fig. 2.12 "Dout" cycle: RE# is low, all other inputs signals are low

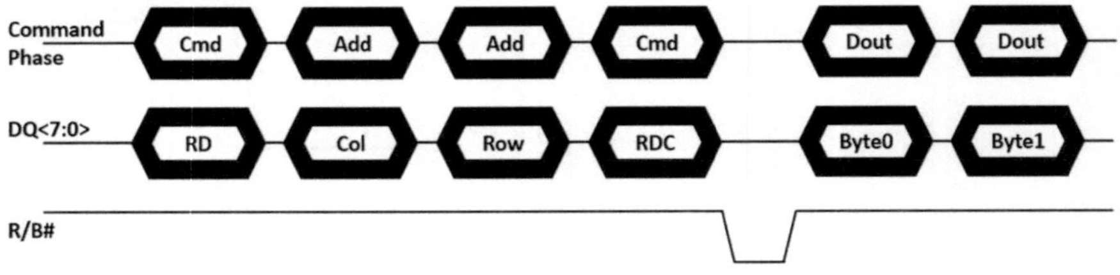

Fig. 2.13 Read command sequence

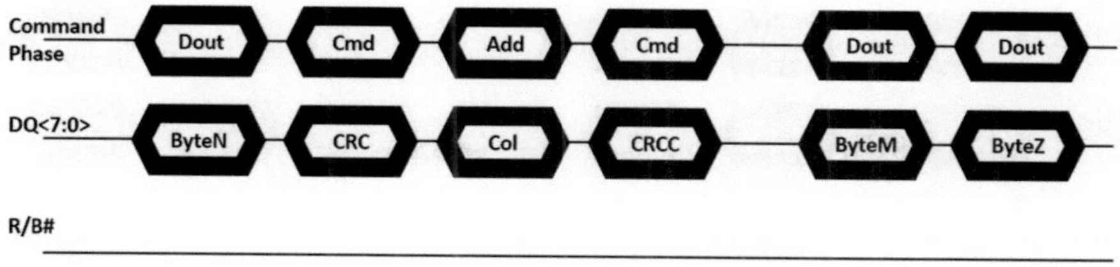

Fig. 2.14 Change read column sequence

to confirm the read command. Finally, the device goes busy and the read operation starts. When NAND returns ready, the data output cycles start.

The above described read command outputs the entire Flash page, regardless the number of bytes we want to read. In some cases, a small data move is required or we may want to read randomly inside a page. Command Change Read Column, also known as Random Data Output, is designed to change the column address during data out.

Figure 2.14 shows the Change Read Column sequence. After the usual read command is executed, it is possible to change the column address during data out. A command cycle "CRC" (Change Read Column, e.g. 05h) is issued, followed by the addresses of the locations we want to output from. Only fewer cycles are required with respect to the usual read command, since only the column addresses are needed. A confirm command cycle "CRCC" (Change Read Column Confirm, e.g. E0h) is used to enable the data out. It is worth noting that no additional busy time is necessary, because data are already stored in the page buffers.

Generally, the busy time in a read operation lasts for several tens of microsecond. One way to improve read throughput is the Read Cache Command (when available). With this command it is possible to download data from the Flash memory, while page buffers are reading another page from the Flash array.

Sequential Read Cache Command sequence is shown in Fig. 2.15. A Read Command must be issued before Read Cache Command. After the device returns ready, the command code "RC" (Read Cache, e.g. 31h) is used to initiate data download from the matrix to page buffers. RB# goes low for a little while and then N Dout cycles are used to output the first page. Since no other addresses are input, the next

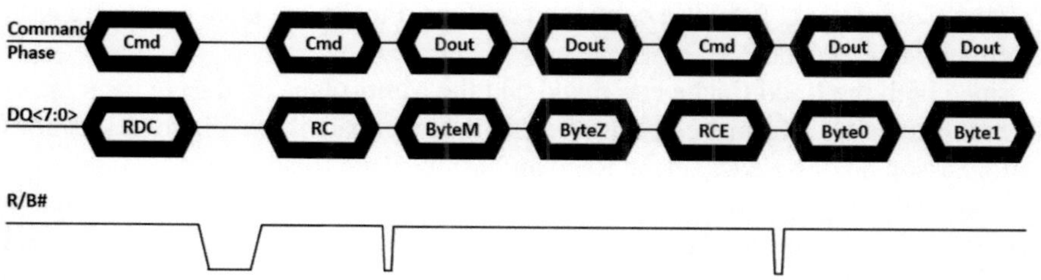

Fig. 2.15 Sequential cache read command

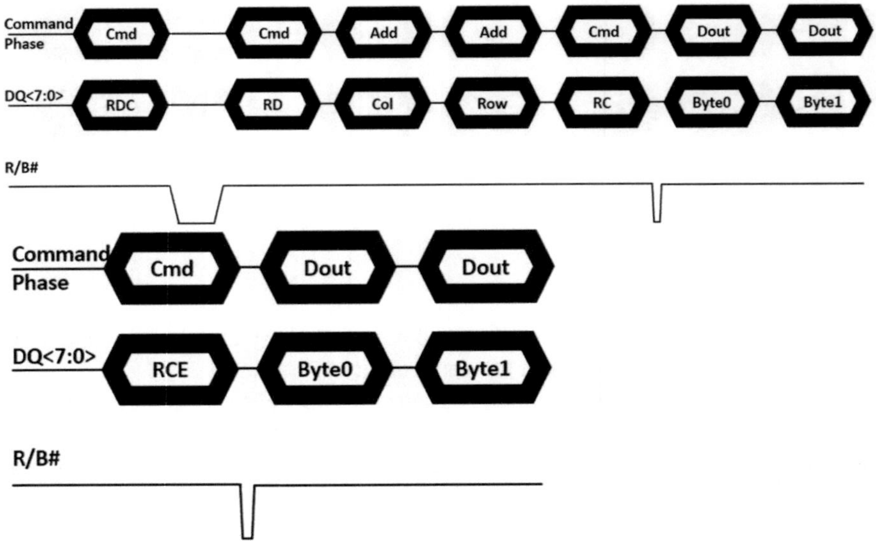

Fig. 2.16 Random cache read command

sequential page is automatically read inside the NAND. When we don't need to read other pages, the last page is copied into the page buffers by using command "RCE" (Read Cache End, e.g. 3Fh).

Random Cache Read sequence is shown in Fig. 2.16: with this command it is possible to select the address of the page we want to cache.

With multi-plane devices, it is possible to issue a read command on multiple planes simultaneously. Figure 2.17 shows the command sequence for Multi-plane Read.

After the standard Read Command cycle "RD", the addresses of the page we want to read on plane 0 are issued. The command code "MR" (Multi-plane read, e.g. 32h) is used in the next command cycle so that the device is ready to receive the addresses belonging to plane 1. Once the new addresses and the Read Command Confirm Code "RDC" are given, the device goes busy to perform the read algorithm on both planes simultaneously. When the device returns ready, the command cycle CC (Choose Column, e.g. 06h) is used to select the address of the page we want to output, followed by a number of address cycles. The command code "CCC" (Choose Column Confirm, e.g. E0h) is a command confirm. Finally the Dout cycles are used to output the read data.

Since both the Read Cache command and the Multi-plane read have been introduced to increase performances, it is interesting to compare them. Figure 2.18 shows a comparison among Read, Cache Read and Multi-plane Read.

If we define t_{ALGO} as the time to read from the NAND array, and t_{OUT} the time to download the page, the total time to perform the read of 2 pages with a standard read is $t = 2\ t_{ALGO} + 2\ t_{OUT}$. If a multi-plane command is used, t_{ALGO} runs simultaneously on both planes and $t = t_{ALGO} + 2\ t_{OUT}$. Evaluating t in the Cache Read case is more

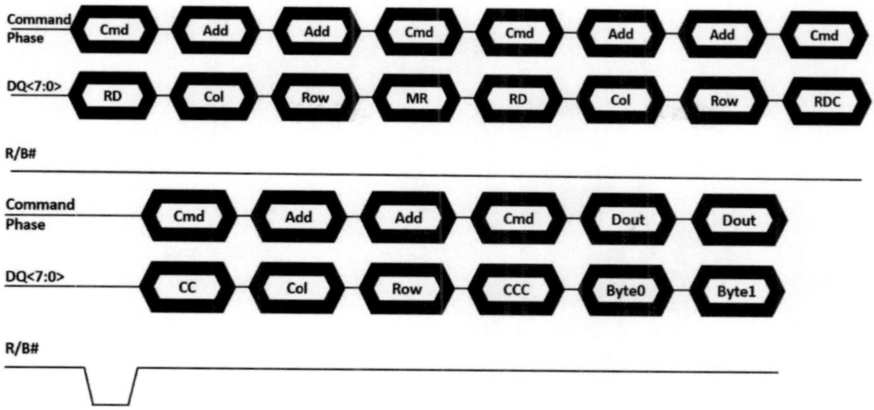

Fig. 2.17 Multi-plane read command

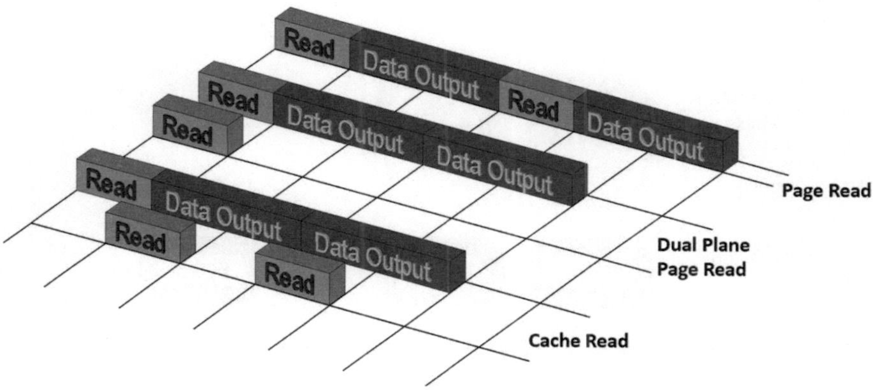

Fig. 2.18 Performance comparison among read, double-plane read and cache read

difficult, because we have to consider the ratio between t_{ALGO} and t_{OUT}. If t_{OUT} is longer than t_{ALGO}, then $t = t_{ALGO} + 2\ t_{OUT}$. It follows that the performances of Cache Read and Double Plane Read are the same. On the contrary, if t_{ALGO} is longer than t_{OUT}, it won't be possible to mask t_{ALGO} with a single page data out (Fig. 2.19). In this case, Double-plane Read performs better than Cache Read.

2.5.2 *Program Operation*

Purpose of program operation is to write data at a specified address. The basic cycles are those already described for read operation, such as Command cycle and Address cycle. The only added cycle is the Data in ("Din") cycle, as sketched in Fig. 2.20.

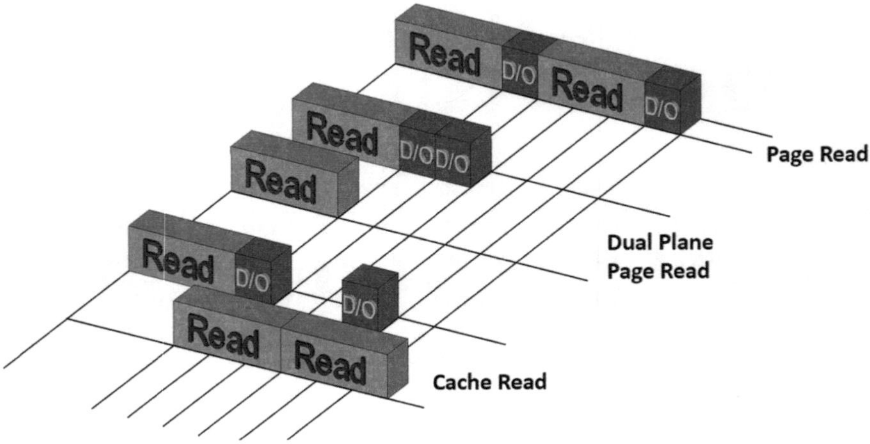

Fig. 2.19 Performance comparison among read, double-plane read and cache read with a NAND array read time longer than page data output

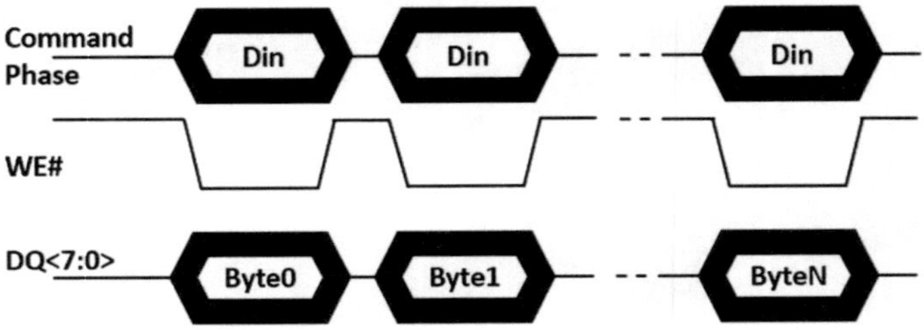

Fig. 2.20 "Din" cycle: WE# is low, all other inputs signals are low

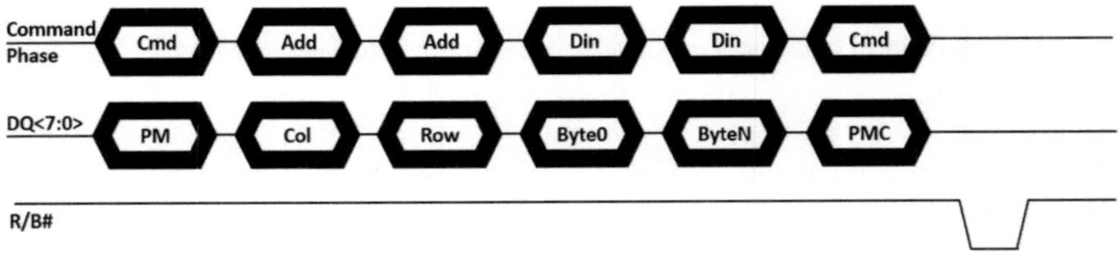

Fig. 2.21 Program command

Data in is performed by toggling signal WE#: at every cycle a new byte shall be made available on DQs.

Program sequence is shown in Fig. 2.21. A command cycle to input "PM" code (Program, e.g. 80h) is followed by a number of address cycles to input the addresses where we want to write. Once the location is set, N "Din" cycles are used to input data into the page buffers. Finally a "PMC" (Program Confirm, e.g. 10h) command is issued to start the algorithm.

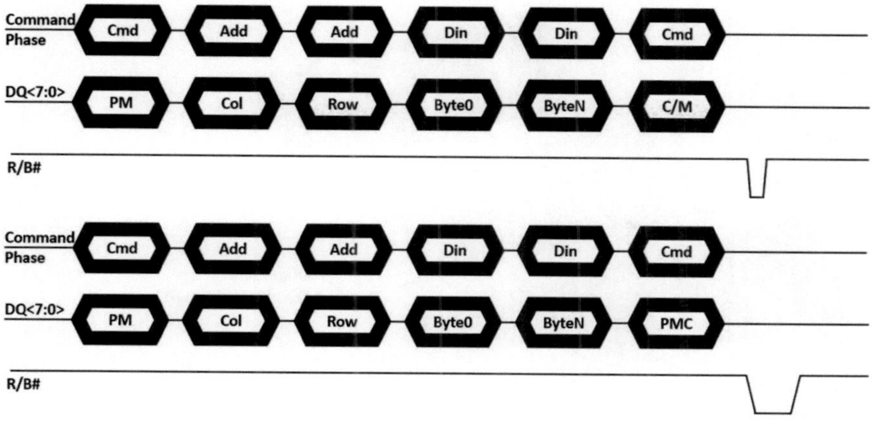

Fig. 2.22 Cache program and double-plane program commands

As already described for read operation, also in the program case there could be the need to move a small amount of data. Change Write Column ("CWC") is used to change the column address where we want to load the data.

Program busy time can be as long as few millisecond. Program cache command or double-plane program are used to increase write throughput. Figure 2.22 shows the sequence for Cache Program and Double Plane Program.

The first cycles ("PM" cycle, address cycles and "Din" cycles) are the same as in the standard program. Instead of "PMC" a "C/M" command cycle is issued. "C/M" can be the Cache Program Code (e.g. 15h) or a Double Plane Command (e.g. 11h). Once another "PM" command is given, followed by the new addresses and the "PMC" command, the device goes busy and the program algorithm is performed simultaneously on both pages. It is worth noting that the above described Double plane program is generally known as Concurrent double-plane Program, because the program algorithm works simultaneously on both planes.

Overlapped Double-Plane Program might also be available; in this case, programming of the first plane starts as soon as data are loaded in the page buffers. Of course, this functionality requires a NAND architecture capable of performing the programming algorithm independently on both planes.

The comparison between the above mentioned program commands is shown in Fig. 2.23.

2.5.3 *Erase Operation*

The Erase Operation is used to delete data from the Flash array. Figure 2.24 shows the Erase Command sequence.

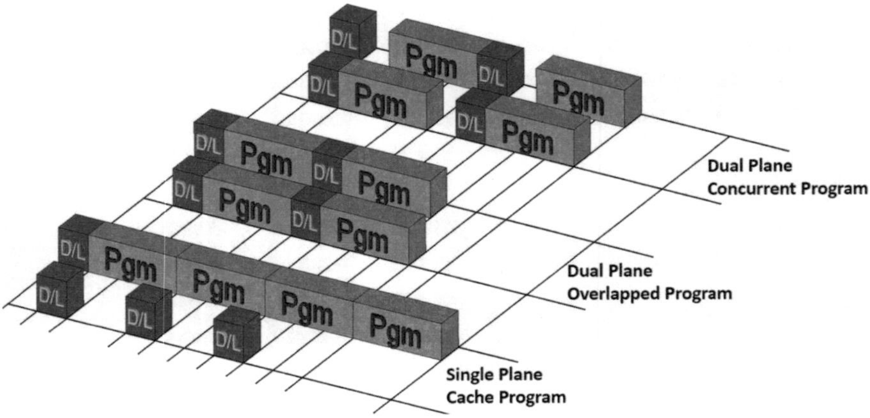

Fig. 2.23 Performance comparison among cache program, overlapped double plane program and concurrent double plane program

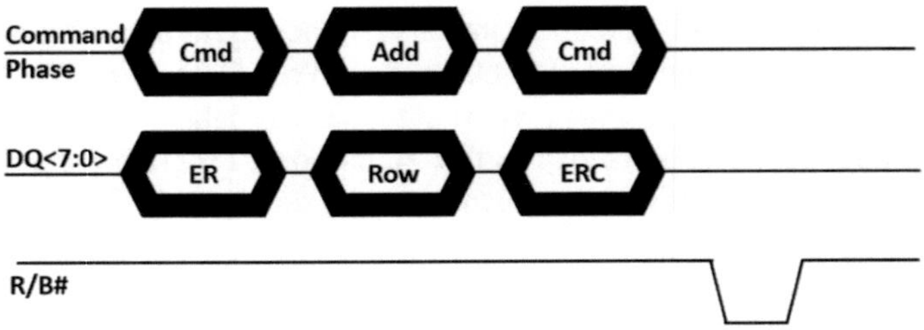

Fig. 2.24 Erase command

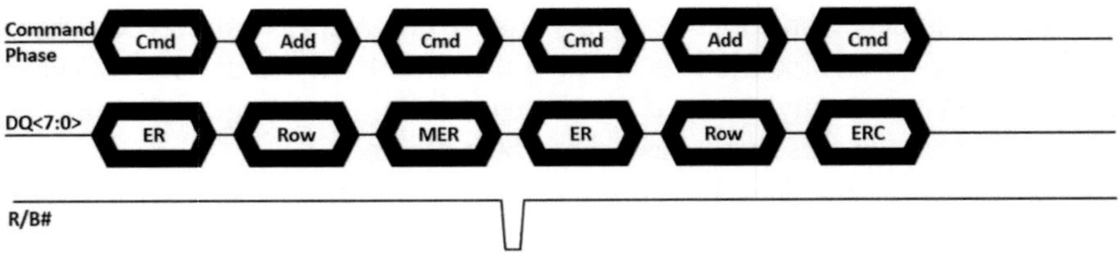

Fig. 2.25 Double-plane erase command

The Erase command is very simple: "ER" code (Erase Command, e.g. 60h), followed by the block address and the "ERC" code (Erase Command Confirm, e.g. D0h). After that, the device goes busy to perform the algorithm.

Since erase is the longest operation, the Double-Plane Erase command has been introduced to erase two blocks at the same time. Figure 2.25 shows the command sequence for the Double-Plane Erase.

The standard erase cycles ("ER" command and row address cycles) are followed by a "MER" command (Multi-plane erase, e.g. D1h). Once both the plane 1 addresses and the "ERC" code are given, the device goes busy, erasing both blocks simultaneously.

2.6 Synchronous Operations

NAND read throughput is determined by array access time and data transfer rate over the DQ bus. Typically, the data transfer is limited to 50 MB/s by the asynchronous interface. As technology shrinks, page size increases and data transfer takes longer; as a consequence, NAND read throughput decreases, totally unbalancing the ratio between array access time and data transfer on the DQ bus. DDR interface has been introduced to balance this ratio [8].

Nowadays two possible solutions are available in the market. The first one, *Source Synchronous Interface* (SSI), is driven by the ONFI (*Open NAND Flash Interface*) organization established in 2006 with the purpose of standardizing the NAND interface [9]. Other NAND vendors use the so-called Toggle-Mode interface. In 2012 JEDEC defined a NAND flash device interface interoperability standard (JESD230) that supports Asynchronous, SSI and Toggle DDR [10].

Figure 2.26 shows the NAND pinout for SSI. Compared to the *Asynchronous Interface* (ASI), there are 3 main differences:

- RE# becomes W/R# which is the Write/Read direction pin;
- WE# becomes CLK which is the clock signal;
- DQS is an additional pin acting as the data strobe, i.e. it indicates the data valid window.

Hence, the clock is used to indicate where command and addresses should be latched, while a data strobe signal is used to indicate where data should be latched. DQS is a bi-directional pin and it runs at the same frequency of the clock.

Obviously, the basic command cycles described in the previous sections must be modified to comply with the synchronous interface.

For example, Fig. 2.27 shows a "Cmd" sequence, followed by Data out ("Dout") cycles for SSI.

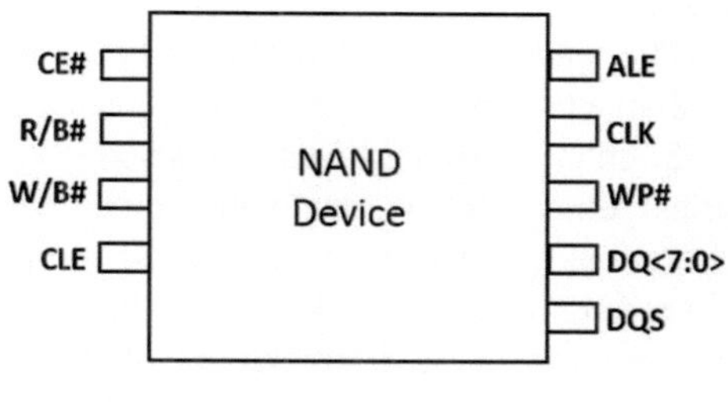

Fig. 2.26 Pinout of a NAND Flash supporting source synchronous interface

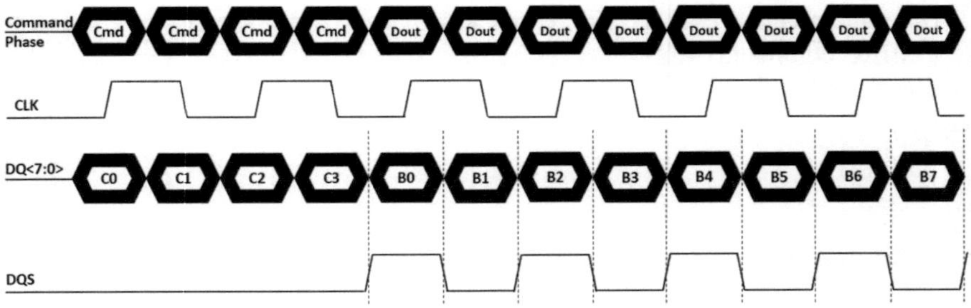

Fig. 2.27 Source synchronous interface DDR sequence

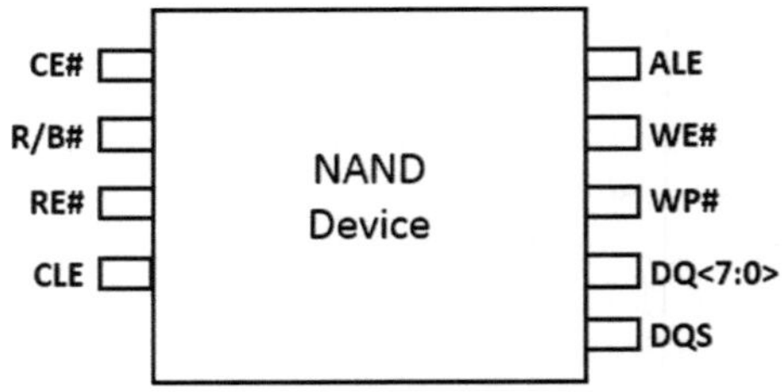

Fig. 2.28 Pinout of a NAND Flash supporting toggle-mode interface

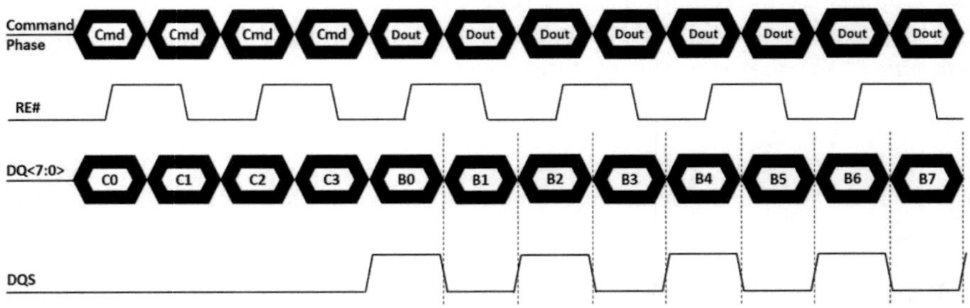

Fig. 2.29 Toggle-mode DDR sequence

Toggle-Mode DDR interface uses the pinout shown in Fig. 2.28.

It can be noted that only the DQS pin has been added to the standard ASI. In this case, higher speeds are achieved by increasing the toggling frequency of RE#.

Figure 2.29 shows a "Cmd" sequence, followed by "Dout" cycles for Toggle-Mode interface.

The reader can refer to NAND Flash datasheets for a detailed description of all available commands with Source Synchronous and Toggle-Mode interfaces.

Simulations of Solid State Drives involve several synchronous NAND Flash memories, some of them doing reads, some others doing either write or erase; moreover, operations can be cached, interleaved, multi-plane, etc., in all possible combinations, depending on the drive workload. This is clearly a non-linear problem and this book is about how to make simulations of such a complex system, by providing insights on available tools and simulation strategies. As usual, speed and precision don't go hand in hand and it is important to understand when to simulate what, and with which tool.

References

1. G. Campardo, R. Micheloni, D. Novosel, "VLSI-Design of Non-Volatile Memories", Springer-Verlag, 2005.
2. R. Micheloni, "3D Flash Memories", Springer, 2016.
3. R. Micheloni, L. Crippa, A. marelli,"Inside NAND Flash Memories", Springer, 2010.
4. S. Lee et al., "A 128Gb 2b/cell NAND Flash Memory in 14nm Technology with t_{prog}=640μs and 800Mb/s I/O Rate", 2016 IEEE International Solid-State Circuits Conference (ISSCC), Dig. Tech. Papers, pp. 138–139, San Francisco, USA, Feb. 2016.
5. R. H. Fowler and L. Nordheim, "Electron Emission in Intense electric Fields," Proceedings of the Royal Society of London, vol. 119, no. 781, pp. 173–181, May 1928.
6. R. Micheloni et al., "A 4Gb 2b/cell NAND Flash Memory with Embedded 5b BCH ECC for 36MB/s System Read Throughput", IEEE International Solid-State Circuits Conference Dig. Tech. Papers, pp. 142–143, Feb. 2006.
7. R. Micheloni, A. Marelli, R. Ravasio, "Error Correction Codes for Non-Volatile Memories", Springer-Verlag, 2008.
8. D. Nobunaga et al., "A 50nm 8Gb NAND Flash Memory with 100MB/s Program Throughput and 200MB/s DDR Interface", IEEE International Solid-State Circuits Conference Dig. Tech. Papers, pp. 426–427, Feb. 2008.
9. www.onfi.org
10. www.jedec.org

Chapter 3
SSDExplorer: A Virtual Platform for SSD Simulations

Lorenzo Zuolo, Cristian Zambelli, Rino Micheloni and Piero Olivo

Abstract *Solid State Drives* (SSDs) are becoming more and more popular, driven by the restless growth of high performance computing and cloud applications. The development of an SSD architecture implies the analysis of a bunch of trade-offs that, if properly understood, can tighten the SSD design space, thus reducing the prototyping effort. Although SSD hardware prototyping platforms are the best way to capture realistic system behaviors, they inherently suffer from a lack of flexibility. To tackle this challenge and to identify the optimum design, under a given set of constraints, the SSD research community is increasingly relying on sophisticated software tools for modeling and simulating SSD platforms. In the first part of this chapter the authors take a careful look at both literature and available simulation tools, including VSSIM , NANDFlashSim, and DiskSim. All these solutions are benchmarked against performances of real SSDs, including an OCZ VERTEX 120 GB and a NVRAM card used in large enterprise storage platform, that have been measured under different traffic workloads. PROs and CONs of each simulator are analyzed, pointing out which kind of answers each of them can give and at what price. The second part of the chapter is devoted to an advanced simulator named "SSDExplorer", which is a fine-grained SSD virtual platform that was developed with the following goals in mind:

- offer a RAD tool (Rapid Application Development) for SSD design space exploration;
- accurately predict performance figures of the target architecture;
- offer a wear-out aware framework for complex correction algorithm exploration;
- avoid the overdesign of the resources of the SSD for a target performance.

Thanks to these features, different drive architectures can be compared and modified until the optimum SSD architecture is identified, without the need for any hardware prototyping.

L. Zuolo (✉) · C. Zambelli · P. Olivo
Dipartimento di Ingegneria, Università degli Studi di Ferrara, via G. Saragat, 1, 44122 Ferrara, Italy
e-mail: lorenzo.zuolo@unife.it

R. Micheloni
Performance Storage Business Unit, Microsemi Corporation, Vimercate, Italy

R. Micheloni (ed.), *Solid-State-Drives (SSDs) Modeling*, Springer Series in Advanced Microelectronics 58,
DOI 10.1007/978-3-319-51735-3_3

Solid State Drives (SSDs) are becoming more and more popular, driven by the restless growth of high performance computing and cloud applications [1]. The development of an SSD architecture implies the analysis of a bunch of trade-offs that, if properly understood, can tighten the SSD design space, thus reducing the prototyping effort. Although SSD hardware prototyping platforms are the best way to capture realistic system behaviors, they inherently suffer from a lack of flexibility [2].

To tackle this challenge and to identify the optimum design, under a given set of constraints, the SSD research community is increasingly relying on sophisticated software tools for modeling and simulating SSD platforms. There are basically two categories of tools: drive emulation tools [3] in virtual environments [4] and pure software simulation tools [5]. The former category uses functional simulations to quickly evaluate SSD performances from the host perspective. This comes at the cost of reduced design space exploration capability, mainly because of the higher level of abstraction. The latter category is built on top of trace driven simulators, thus focusing on the analysis of performances when the drive is in a steady-state. Both tools often overlook the macroscopic performance/reliability implications of microarchitecture-level effects, which are typically abstracted.

In both categories of tools the common underlying assumption is that the SSD microarchitectural details that determine the drive behavior are already defined. In practice, the modeling framework does not target a *Fine-Grained Design Space Exploration* (FGDSE) of the microarchitecture; primary goals are performance estimation of known architectures, full system simulation, and *Flash Translation Layer* (FTL) validation. These approaches are in general not very useful for an SSD designer, whose primary need is not the capability of performing functional simulations, but rather the possibility of quantifying the efficiency of microarchitectural design choices to cope with long term concerns, such as drive reliability and wearout-induced performance drops.

SSDExplorer was born to fill the gap in the landscape of simulation frameworks for SSD devices, specifically targeting FGDSE. Bridging this gap is mandatory to avoid the over-design of an SSD architecture when trying to meet target I/O performances and reliability.

Based on a software framework, SSDExplorer complements modeling and simulation capabilities of the state-of-the-art tools. Here is a list of the main features.

- *All components of the SSD architecture are modeled*, thus broadening the design space exploration capabilities. Moreover, a different level of abstraction can be associated to each SSD component.
- *Impact of the FTL does not necessarily require a detailed FTL description.* This is achieved by supporting the *Write Amplification Factor* (WAF) abstraction [6]. As a consequence, one of the goals of SSDExplorer is to deliver a fast path for accurate I/O performance estimation.
- *Accurate performance breakdown and identification of microarchitectural bottlenecks.* Analysis of the interaction efficiency between subcomponents of an SSD is an essential requirement for the microarchitecture design.

- Accuracy of the simulation results were validated with a mature commercial platform (i.e., Ocz Vertex 120 GB [7]) and a state-of-the-art enterprise platform [8].

Thanks to these features, different drive architectures can be compared and modified until the optimum SSD architecture is identified, without the need for any hardware prototyping.

3.1 SSD Simulators

As mentioned above, disk *emulation* [3] and disk *trace-driven* simulation software tools [5, 9, 11–15] have been used for many years.

Yoo et al. [3] proposed a disk *emulation* strategy based on a reconfigurable framework able to deal with a real SSD. One of the key contributions of this work is the ability to track the real performance of a host system through a dynamic manager built around a QEMU virtual platform [4]. However, to achieve fast performance estimations, several components (i.e. processor, NAND Flash memories, etc.) are modeled with a high level of abstraction. Therefore, performance fluctuations experienced by these blocks are lost, thus strongly reducing the performance estimation accuracy.

In the context of SSD *trace-driven* simulation tools, the open-source frameworks proposed in [5, 9] allow SSD performance and power consumption evaluation. Attempts to improve them in order to achieve real performance matching were also proposed in [14, 15]. However, these tools are still highly abstracted, thus providing an insufficient level of simulation accuracy and realistic components description to perform real FGDSE. Moreover, since the aforementioned classes of frameworks do not model all the internal blocks of an SSD, they are able to accurately track the behavior of a disk only starting from a set of predefined and statically assigned timings (i.e. channel switch delay, I/O bus latencies, command scheduler delay, etc.). An additional attempt to modify one of these tools in order to incorporate detailed NAND and ECC timings was done in [16]. Although the accuracy of the obtained results in that particular case study is high, this kind of tools still lack the ability of evaluating micro-architectural effects on SSD performances, like commands pipelining, command suspensions, and uncommon queuing mechanisms, which become visible only with cycle-accuracy.

To overcome this weakness, several cycle-accurate SSD simulators were developed. For example, Lee et al. [11] adopted a precise clock simulation for hardware components description. However, this approach does not allow a full modeling of all SSD components, thus hiding some of the architecture bottlenecks. Other methods for fast simulation were presented in [12, 13], but they also suffer from accuracy loss due to the lack of a complete architectural modeling.

Hardware platform prototypes were described in [2] and [10]. They enable a precise SSD behavior investigation, although their fixed architecture severely limits

Table 3.1 Comparison of the most used SSD simulation frameworks

Reconfigurable parameters	SSDExplorer platform	Emulation platforms (VSSIM [3])	Trace-driven platforms (DiskSim, FlashSim [5, 9])	Hardware platforms (OpenSSD, BlueSSD [2, 10])
Real FTL	√	√	√	√
WAF FTL	√	No	No	No
Host interface performance	√	√	No	√
Real workload	√	√	No	√
Different host interfaces	√	No	√	No
Accurate DDR timings	√	No	No	No
Multi DRAM buffer	√	No	No	No
Configurable channel Nº	√	√	√	No
Configurable target Nº	√	√	√	No
NAND architecture	√	√	√	No
Accurate NAND timings	√	No	No	√
NAND reliability	√	No	No	√
ECC model	√	No	No	√
Interconnect model	√	No	No	√
Core model	√	No	No	√
Real firmware execution	√	No	No	√
Multi core	√	No	No	No
Model refinement	√	No	No	No
Simulation speed	Variable	High	High	Fixed

the exploration of different design solutions. In practice, only the internal firmware can be modified.

What is really missing in these approaches is a clear exploration of the performance correlation between the host interface capabilities and the non-volatile memory subsystem, including all the architectural blocks in between.

Table. 3.1 compares the relevant features of the most used SSD simulation frameworks. The main highlight is that SSDExplorer introduces detailed timings and behavioral models of the critical architectural blocks (i.e. DRAMs, NAND memories, ECC sub-system, etc.) that are mandatory for an accurate performance/reliability evaluation.

3.2 SSDExplorer at a Glance

3.2.1 *Modeling Strategy*

One of the main reasons that drove the development of SSDExplorer was the desire to have a unified, reconfigurable and multi-abstraction simulation environment. To achieve this goal, each block of SSDExplorer is written and integrated by using the SystemC modeling and simulation environment [17]. SystemC allows designers to cover, by using a single description language, several level of abstraction, from the *Timed Functional Level* up to the *Register Transfer Level* (RTL). Thanks to this approach, if a specific block needs to be thoroughly investigated, designers can easily plug a more accurate model into the simulation environment, without any impact on other components. However, it is worth highlighting that, since the simulation speed offered by SystemC is inversely proportional to the abstraction level, each block has to be wisely modeled; simulation efficiency must always be tailored to the simulator goals. Although a similar strategy was successfully adopted in other applications [18], this is the first application in a FGDSE-dedicated SSD simulation tool.

In essence, SSDExplorer is designed to: (i) select the most suitable modeling style for each SSD component in order to accurately quantify performances; (ii) tolerate lack of precise implementations of specific HW/SW components by providing modeling abstraction. In fact, a detailed implementation of all SSD components might not be available during the early architectural analyses. Based on the above considerations, all HW/SW components that logically belong to the SSD control path (i.e. all the blocks involved in the command interpretation) are modeled with high accuracy; on the contrary, components on the datapath (i.e. used during the data transfer phase) can be reduced to a simple delay. This approach improves the simulation speed while capturing all those micro-architectural details affecting SSD performances; at the same time, this approach provides a backbone for FTL functional simulations. Communication between model domains is guaranteed through "wrappers", which translate logical signals into state variables.

3.2.2 *Available Models in SSDExplorer*

Figure 3.1 shows the reference SSD architecture simulated by SSDExplorer. There are three levels of abstraction [19]: *Pin-Accurate Cycle-Accurate* (PA-CA), *Transaction-Level Cycle-Accurate* (TL-CA), and *Parametric Time Delay* (PTD).

Pin-Accurate Cycle-Accurate (PA-CA) models

CPU, system interconnect, and channel/target controller are the key components in the data flow. All these blocks must be included when the target is to simulate FTL, either real (if available) or abstracted. FTL is usually implemented at the firmware (FW) level and its overhead might have a significant impact on the overall SSD

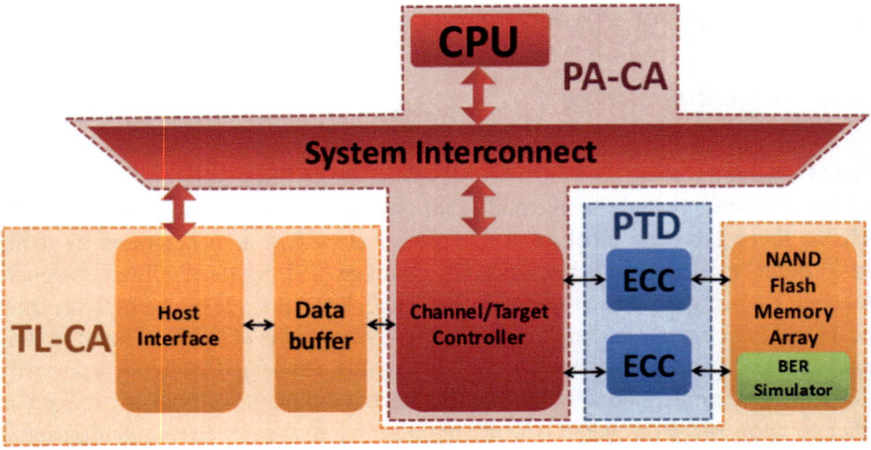

Fig. 3.1 Reference SSD architecture simulated by SSDExplorer

performances. As such, to accurately capture commands and their timings, a cycle-accurate description is required.

Cycle-accurate models can effectively describe the complete functionality, structure, communication, and timing of a micro-architectural block. The communication between components is modeled in a PA-CA manner: the hardware signals connecting a component to a bus are explicitly modeled, and all the timing and protocol-induced delays are captured accurately for the bus architecture. The components are modeled in a cycle-accurate manner as well: data processing inside a component is scheduled on a cycle by cycle basis. The structure of cycle-accurate models is very close to RTL, but it allows faster simulations thanks to a higher-level modeling and to the simulation language.

In SSDExplorer, pin- and cycle-accuracy are mandatory for modeling the control path, since subtle effects in command handling and/or component interactions may cause performance deviations that should be caught by a FGDSE tool. Any end user may plug its SystemC models here, reflecting in-house IP cores. In the early design phases, however, more relaxed bus functional models can be used, limiting cycle accuracy to the bus interfaces and to the bus architecture itself. This option reduces the simulator capability in capturing FTL execution overhead, but it does not limit the drive performance estimation, by using either coarse or abstracted FTL models.

1. CPU: SSDExplorer can implement any CPUs, including custom IP cores and advanced multicore, starting from its PA-CA model, its proper Instruction Set Simulator, or the actual processor command back-trace. Thanks to the built-in CPU-wrapper, users can even design and optimize custom CPUs for FTL execution by using other tools; once the set of operations is defined, it can be easily integrated into SSDExplorer. Of course, with such a flexibility, it becomes easy to support the inter-operability with state-of-the-art processors simulators such as Gem5 [20] and Open Virtual Platforms [21].

2. System Interconnect: SSDExplorer can include the most relevant communication interfaces used in SSD platforms such as: AMBA AHB, Multi-Layer AMBA AHB, AMBA AXI (single address lane, multiple data lanes) and OCP. Custom system interconnects can also be plugged into the simulator if their PA-CA models are available.
3. Channel/Target Controller: to perform read/write operations from/to the NAND Flash memory arrays, it is mandatory to have a specific controller for translating the commands issued by the CPU into a low-level NAND commands. The *Open NAND Flash Interface* (ONFI) [22] and *Toggle* are the most widely used NAND Flash Interface standards as we speak [23]. From an architectural point of view, the channel/target controller is made of five macro blocks: a slave program port on the system interconnect, a Push-Pull DMA controller, a SRAM cache buffer, an ONFI-Toggle port and a command translator. The microarchitecture described in [24] was chosen to mimic realistic functionalities of a channel/target controller in industry-relevant designs, and it is shown in Fig. 3.2. SSDExplorer can be configured with a flexible number of channels and targets.

Transaction-Level Cycle-Accurate (TL-CA) models
Host interface, DRAM buffers and NAND Flash memory arrays are described by selectively abstracting the modeling style. The main idea is to avoid modeling all pins of the communication interface between the data path components, as well as the signals that constitute the bus and the external system interface. Basically, communications go through channels, on top of which read and write transactions are carried out. At the same time, computation primitives are scheduled at each clock cycle. This is to allow the execution of sudden command requests in parallel with other transactions (e.g. erase suspend in NAND Flash memories during garbage

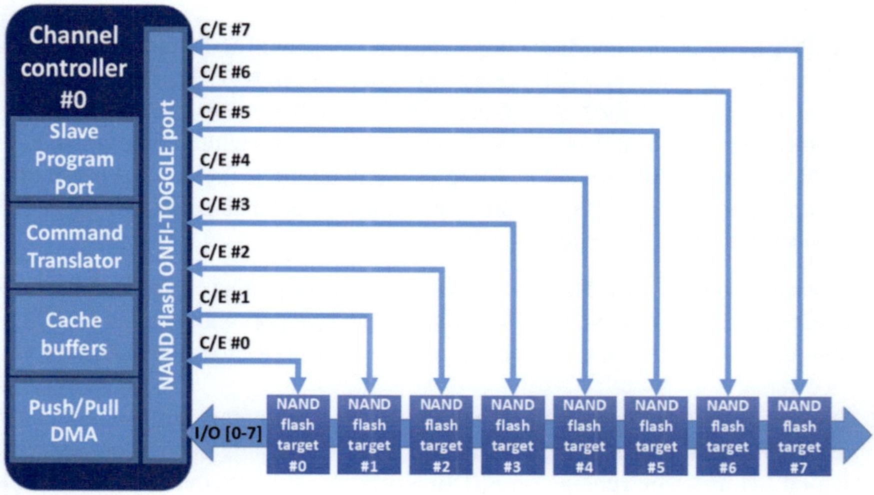

Fig. 3.2 Architecture of the channel/target controller

collection), and to preserve timing accuracy in the wrappers bridging these models with the pin- and cycle-accurate ones. Nevertheless, the burden to preserve this cycle accuracy is not heavy. In fact, there are memory dominated-components on the data path, whose performance mainly depends on properly capturing timing parameters of the memories rather than on the modeling of complex computation tasks. Moreover, since NAND Flash memory components could be inactive for long times, (e.g. for random workloads which do not spread across all the SSD channels), the processes used for memory simulations are triggered only on demand, such that the overall simulation efficiency is maximized. Basically, when the command scheduler detects that an operation must be issued to one or more NAND Flash targets belonging to a single channel, it spawns the corresponding process responsible for the management of the targets of that channel. If the channel is idle, no process is created. Upon process spawning, only the finite state machines of the addressed targets are updated, while the other ones remain idle. By combining the dynamic process management with the selective process spawning, it is possible to mitigate the impact of the memory subsystem on the simulation speed.

1. Host Interface: this component manages the communication protocol with the host, providing commands and data to the SSD. Two types of interfaces are implemented in SSDExplorer: *Serial Advanced Technology Attachment* (SATA) and *PCI-Express*. Both interfaces include a command/data trace player, which parses a file containing the operations to be performed and triggers operations for the other SSD components accordingly. The features of the available interfaces are summarized below.
    - SATA: all SATA protocol layers [25] and operation timings are accurately modeled according to the SATA protocol timing directives provided in [26]. Native Command Queuing support is implemented featuring arbitrary queue length up to 32 commands.
    - PCI-Express: this interface allows boosting sequential and random operations throughput, and it is widely adopted in enterprise SSDs [1]. Fast operations are achieved through the NVM-Express (*Non Volatile Memory Express* [27]) protocol that significantly reduces packetization latencies with respect to standard SATA interfaces [28]. All PCI-Express configurations (i.e. from generation 1 up to generation 3 with variable lane numbers) can be modeled, thus ensuring accurate latency matching.

    To easily switch from one host interface to another, a common control architecture is adopted: it is based on a fabric interconnect slave port and an external DMA controller [29] able to transfer data from the host interface to the data buffers and vice versa.
2. DRAM Buffer: this component is used either as a temporary storage buffer for read/write data or as a buffer [31] for the address mapping operations, depending on the firmware running on SSDExplorer. A cycle accurate DRAM model is required to capture realistic behaviors (i.e. column pre-charging, refresh operations, detailed command timings, etc.). The data buffers of SSDExplorer are

modeled with a SystemC customized version of the simulator proposed in [32]. The number of available buffers in a SSD architecture is upper bounded by the number of channels served by the drive controller. In SSDExplorer the user can freely change this number, as well as the bandwidth of the memory interface, the DRAM functionality, etc., through a simple text configuration file, which abstracts internal modeling details. DDR, DDR2, DDR3 and DDR4 protocols are supported as DRAM interface. Figure 3.3 shows an example of the file used to configure the timings of the DDR Buffer.

3. NAND flash memory array: the fundamental component of an SSD is the non-volatile storage memory array. NAND Flash devices are hierarchically organized in dies, planes, blocks, and pages. Program and read operations work on a page basis, whereas the erase operation is performed per block, thus inhibiting the in-place data update. Due to the internal architecture of NAND Flash devices, memory timings strongly depend upon the selected operation; therefore, there is a significant performance variability [33]. To accurately take all these effects into account, a cycle accurate NAND Flash simulator is used. Furthermore, to emulate the realistic behavior of the memory array, an error injection engine (i.e. a RBER simulator) is included; in this way, it is possible to evaluate the impact of different error patterns on the other components of the NAND Flash/ECC subsystems. Of course, it is possible to embody different NAND Flash technologies (i.e., single-, multi- and triple-level cell). Figure 3.4 shows an example of the file used to configure the timings of a NAND flash die.

Parametric Time Delay (PTD) model

The microarchitectural blocks related to *Error Correction Codes* (ECCs) are modeled by using a parametric time delay abstraction level. These blocks strictly depend upon the design choices of SSD vendors, but their behavior and impact on I/O performances can be easily abstracted by means of well-defined quality metrics [34]. In other words, the behavior inside PTD models does not need to be cycle accurate. As a result, computation primitives inside a component can be grouped together and described by a single timing. For example, the correction time of 5 errors in a NAND Flash page could take an effective wait time of 10 μs that cannot be interrupted by any other commands. The above consideration allows a simpler block description, shortening modeling time and increasing simulation speed. At the same time, communication events can still be scheduled in a cycle accurate manner. However, even if this choice enables accurate I/O performance characterization, it is not well suited for functional simulations. This is actually not a big problem in the initial design stage, when the internal SSD architecture is still under definition: priority number one is to find the right architecture for the target I/O performances. Of course, later on, PTD models can be replaced by more accurate models.

SSDExplorer embodies two configurable PTD ECC models: BCH (*Bose, Chaudhuri, Hocquenghem*) and LDPC (*Low Density Parity Check*). As sketched in Fig. 3.5, these blocks are made of a fixed high-speed encoder for each SSD channel and a reconfigurable parallel decoder (i.e. a multi-engine decoder) whose HW engines are shared among the channels. The delays of both the encoder and the decoder can be

Fig. 3.3 Example of DRAM buffer configuration file [30]

```
NUM_BANKS=8
NUM_ROWS=32768
NUM_COLS=512
DEVICE_WIDTH=8
REFRESH_PERIOD=70200

tCK=1.25
CL=11
AL=0
BL=8
tRAS=28
tRCD=11
tRRD=5
tRC=39
tRP=11
tCCD=4
tRTP=4
tWTR=4
tWR=15
tRFC=88
tFAW=24
tCKE=5
tXP=6

IDD0=42
IDD1=52
IDD2P=12
IDD2Q=22
IDD2N=23
IDD3N=35
IDD4W=99
IDD4R=96
IDD5=182
IDD6=12
IDD7=163

Vdd=1.5
```

configured to mimic either a single- or a multi-threaded architecture, as shown in [34]. To explore different ECC architectures, the internal parallelism of each machine can be selected by the user.

Fig. 3.4 Example of NAND flash die configuration file

```
[TIME]
tADL=70
tCS=20
tDH=1
tDS=1
tWC=25
tWP=11
tR=61200
tRC=10
tREA=3
tRR=20
tRST=5000
tWB=100
tWHR=80
tBERS=6000000
tDBSY=500
tPROG=66200
tFEAT=1000

[TYPMAXTIME]
tPROGU=2162000
tRU=76200
tDBSY=1000
tBERS=25000000

[SYS]
NUMS_PLANE=2
NUMS_DIE=1
NUMS_BLOCKS=2096
NUMS_PAGES=512
NUMS_PGSIZE=16384
NUMS_SPARESIZE=1216
NUMS_IOPINS=8
```

Finally, it is worth pointing out that, in other state-of-the-art SSD simulators, ECC is usually neglected. However, an accurate evaluation of SSD performances must take the latency introduced by ECC encoding and decoding into account, especially when looking at performance-reliability trade-offs related to NAND [35].

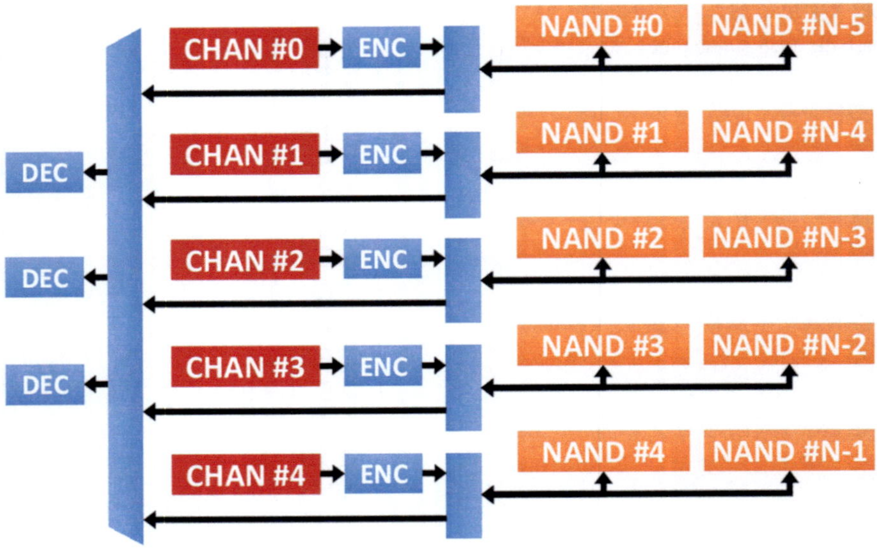

Fig. 3.5 ECC template architecture simulated by SSDExplorer

3.3 FTL Simulations

3.3.1 FTL

During the initial phase of FW development, being able to simulate FTL is critical. Indeed, this is the only way for understanding how Flash management algorithms such as *Garbage Collection* (GC) and *Wear Leveling* (WL) have to be designed, in order to maximize both the performance and the reliability of NAND flash memories. However, since many different implementations are usually available, it is mandatory to offer a flexible framework for both processor simulation/design and FTL execution. In SSDExplorer these problems were addressed by means of two well established open-source simulators: Gem5 [20] and *Open-Virtual-Platform* (OVP) [21]. These tools allow users to define a custom system-on-chip architecture, which can range from single to multi core platforms. Moreover, thanks to the standard programming model offered by these simulators, specific FTL implementations can be easily developed, tested and simulated; in other words, designers can see the actual SSD behavior when the FW runs on top of a specific processor architecture.

All the aforementioned features, however, do not come for free. In fact, about the simulation speed of Gem5 and OVP, it is necessary to highlight that simulations of complex architectures, such as state-of-the-art SSD's processors, can take a long of time. As a consequence, embodying these platforms inside SSDExplorer does not represent the best solution when a quick performance assessment is required. To overcome this problem, there is a dedicated wrapper (sketched in Fig. 3.6) inside the

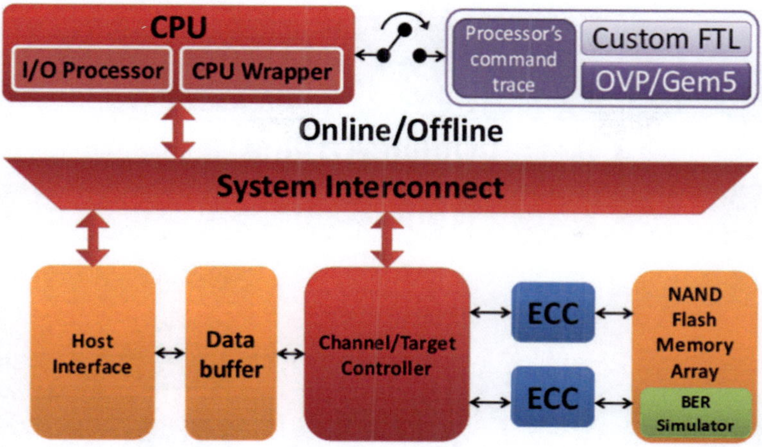

Fig. 3.6 "Online-Offline" FTL simulations

CPU model of SSDExplorer, which can be dynamically attached/detached to/from the processor simulator: this is referred to as "Online-Offline" simulation mode. Basically, the processor simulator runs outside SSDExplorer and it outputs a trace of the commands during the FTL execution ("Offline" simulation). This command trace is composed of two parts: (i) the list of read, write, and erase operations that have to be executed on flash memories; (ii) the processing time taken by the processor to produce each I/O. All the collected data are then fed into the SSDExplorer CPU wrapper ("Online" simulation), which introduces the right amount of delay and sends the I/Os to the I/O processor connected to the Channel/Target controller and the underlying memory arrays. This approach enables simulations of a huge amount of FTL algorithms; in fact, in this case, SSDExplorer plays the role of a simple delay generator, completely agnostic with respect to the actual FTL state or configuration. Moreover, thanks to the "Online-Offline" simulation approach, designers can assess SSD performances only when a specific FTL state has to be studied.

3.3.2 *WAF Model*

The FGDSE of a wide range of SSD architectures has the drawback of requiring a custom FTL tailored for each configuration (some examples of custom FTLs are provided in [36, 36, 37, 37, 38]). Moreover, during the early stages of the SSD design, a complete FTL implementation is not available since many architectural details, such as the processor's architecture, are still under definition. In this context, there is a need for estimating the impact of software management algorithms (e.g. GC, WL, etc.) without requiring their actual implementation. This problem was tackled in [6] by introducing a lightweight algorithm able to evaluate the impact on the SSD's

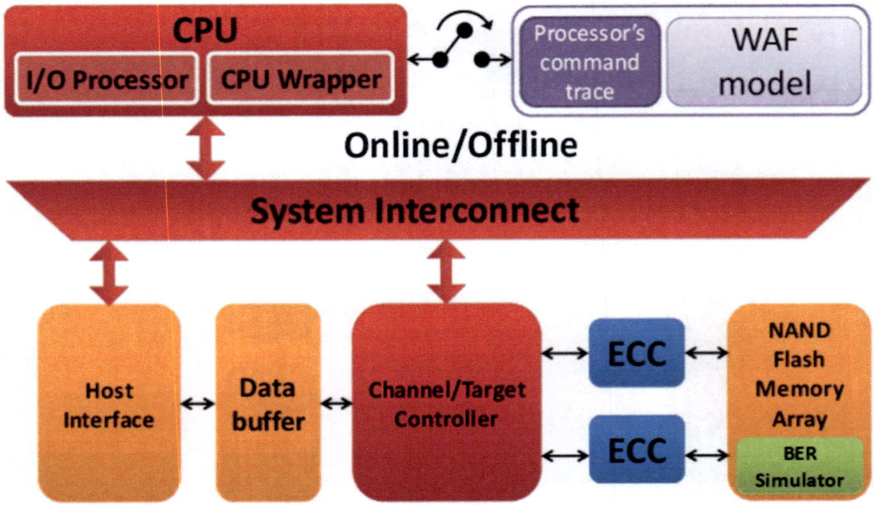

Fig. 3.7 "Online-Offline" FTL simulations with WAF model

performance produced by the GC in terms of the so-called WAF (*Write Amplification Factor*). The algorithm works under the assumption that other FTL functionalities are handled by the CPU without causing a significant performance drop. A standard WAF model [6] computes the number of additional writes caused by the GC operation with respect to the number of writes issued by the host system. Few inputs are required: the total number of blocks in the drive, the over-provisioning factor, and the GC activation threshold which is defined as the percentage of remaining free blocks in the SSD before GC triggers. By using the computed WAF value, it is possible to quickly explore the SSD FTL behavior and assess its efficiency. Generally speaking, the higher the WAF, the longer the GC blocking time, with an heavy impact on the overall drive performances. Thanks to the CPU wrapper and the "Online-Offline" simulation mode, the command trace produced by a WAF model can be input to SSDExplorer, as sketched in Fig. 3.7. Since the target is not a processor simulator, the command trace includes only NAND read, write and erase operations, whereas the actual processing time of each command is neglected.

3.3.3 FTL Versus WAF Model

In order to understand when the WAF model can be reliably adopted, a comparison with a real page-level FTL is required. The architecture sketched in Fig. 3.8 can be used for this exercise. Details of this system are reported in Table 3.2. In this case SSDExplorer is configured with:

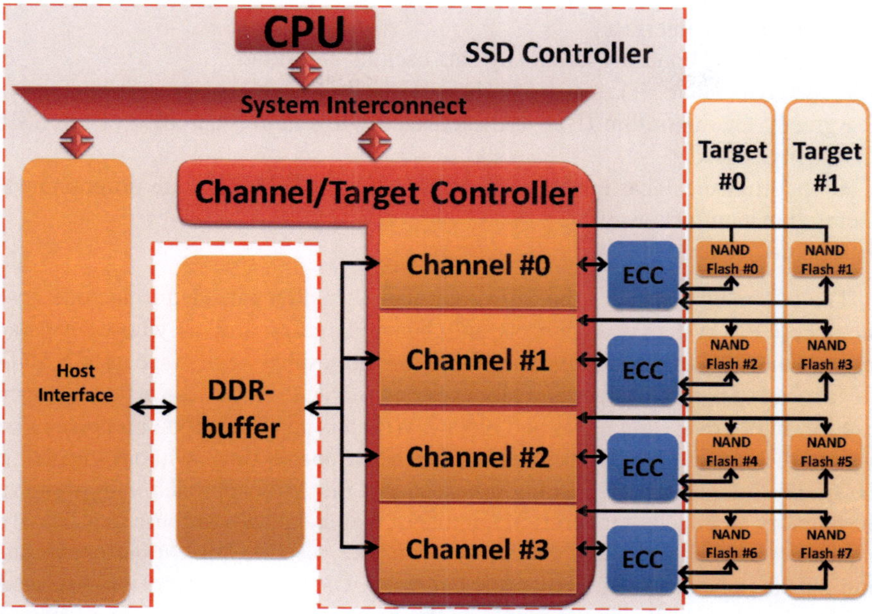

Fig. 3.8 Reference SSD architecture used for comparing FTL and WAF model

Table 3.2 SSDExplorer configuration for WAF accuracy assessment

Parameter	Architecture
Host interface	SATA II
DRAM-buffer	1
Mock-up DRAM-size	64 kBytes
Channels	4
Targets	2
NAND flash dies	4
NAND flash planes	2
NAND flash blocks	16
NAND flash pages	4
NAND flash page size	4096 Bytes
Caching	No
FTL-LOG2PHY mapping	Page-associative
FTL-GC algorithm	Greedy
FTL-GC threshold	70%
FTL-GC reserved blocks	1
FTL-WL policy	Opportunistic
Over-provisioning	20%

- a SATA II host interface;
- a single DRAM buffer without any data caching algorithm;
- a page-associative FTL mapping scheme (FTL-LOG2PHY);
- a greedy GC algorithm (FTL-GC) activated when more than 70% of the SSD capacity is used;
- an opportunistic Wear Leveling approach (i.e. all SSD blocks are programmed/erased in a uniform way);
- over-provisioning of 20%.

The two main actors are the SSD controller (hereafter intended to include host interface, CPU, system interconnects, and channel/target controller), whose key parameter is the CPU frequency, and the NAND memory subsystem. Since each NAND die is made of multiple pages and blocks, it is possible to reduce the actual memory size by simply removing pages and blocks. With this approach, long processes like writing the entire drive can be simulated in a reasonable time, without impacting the framework accuracy. The only caveat to this mock-up approach is to preserve the memory architecture in terms of number of dies and planes because they heavily impact performances. Finally, because of its nature, also the DRAM buffer simulation can take advantages of this approach since the DRAM architecture is constituted by a repetition of banks and channels. In this exercise the buffer size was shrunk to 64 kByte and it is defined by the mock-up DRAM size parameter shown in Table 3.2.

Both FTL and its corresponding WAF model were tested at different SSD controller frequencies, with different workloads. Figure 3.9 shows the performance achieved with a read workload: since read transactions do not require software manipulations, the FTL execution time becomes marginal, thus not affecting the overall bandwidth. When a write workload is considered, Fig. 3.10 shows a discrepancy between the WAF model and the real FTL implementation, which vanishes as the SSD controller frequency goes up. The mismatch at low frequency is mainly due to:

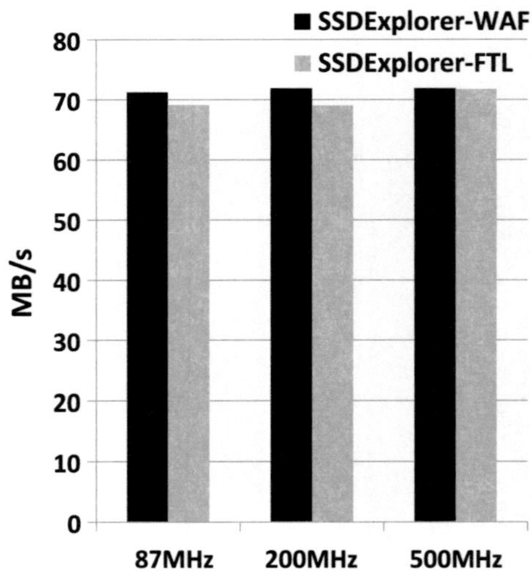

Fig. 3.9 SSDEXplorer simulations of the sequential read bandwidth with FTL and WAF

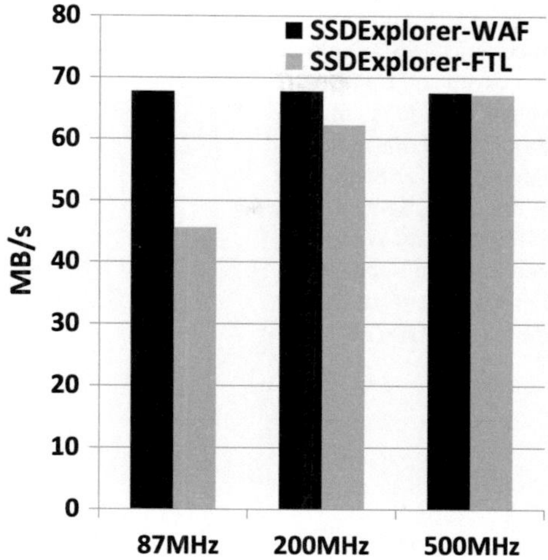

Fig. 3.10 SSDEXplorer simulations of the sequential write bandwidth with FTL and WAF

- the WAF value computed by the model is 1.20, while FTL gives 1.16;
- the additional execution time spent by the FW to identify the GC victim-blocks.

By increasing the SSD controller frequency, execution of the GC management becomes faster; therefore, the FW processing time becomes marginal, and the maximum achievable performances are mainly by the NAND flash timings.

Looking at the above results, it is fair to state that the WAF model can be adopted with a marginal performance misalignment; the only constraint is to stay at a relatively high SSD controller frequency. However, since state-of-the-art SSD controller frequencies are in the 300–600 MHz range [39], the WAF abstraction represents a good speed/accuracy trade-off for FTL simulations.

3.4 Performance Comparison with Real SSD Platforms

3.4.1 Consumer SSD

In order to test the accuracy of SSDExplorer, a direct comparison with OCZ Vertex 120 GB [7], a widely adopted SSD, was carried out. This device was chosen to speed up the validation phase, since it is based on a well-known and documented controller [40], running at 166 MHz, which can be easily simulated. The validation methodology followed in this section makes use of standard synthetic workloads to quantify the I/O performance of SSDS [41]: sequential and random 100% write and 100% read workloads with a block size of 4 kB are injected into the simulated drive. The choice of using synthetic workloads rather than realistic ones [42] is justified by the fact that the latter approach could mask the SSD behavior since the chosen workload may put the SSD in a favorable working point, thus neglecting worst case

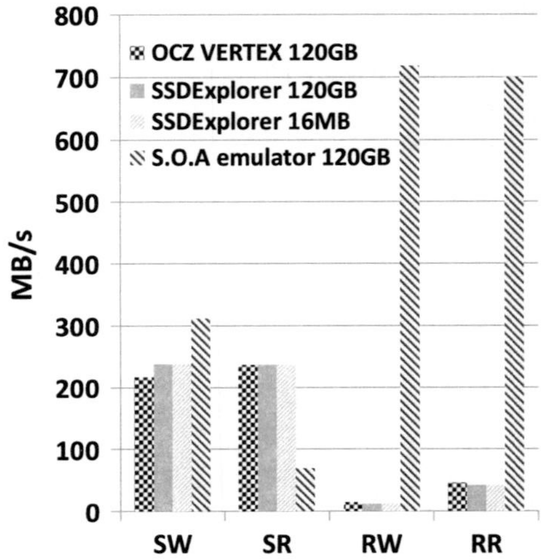

Fig. 3.11 Simulation accuracy of SSDExplorer (mock-up and full drive) versus the S.O.A emulator tool [3] in terms of throughput for Sequential Write (SW), Sequential Read (SR), Random Write (RW) and Random Read (RR). Device under test: OCZ Vertex 120 GB (SSD)

conditions. All the following simulations are based on the WAF abstraction model and they include a full drive of 120 GB and a mock-up version of 16 MB.

As shown in Fig. 3.11, with a sequential workload, SSDExplorer matches the OCZ device performances with an error of about 8% during write and 0.1% during read. With a random workload, the performance deviation from the OCZ drive is 6% for write and 2% for read. These deviations are due to the fact that the WAF model doesn't include any write caching algorithms because details were not published [6]. According to the OCZ Vertex reference manual [7], caching is massively adopted to reduce the amount of write operations redirected to the non-volatile memory subsystem; therefore, simulated write operations (both sequential and random) show offsets higher than read operations. In light of this consideration, results of Fig. 3.11 confirm the accuracy of SSDExplorer. It is worth reiterating the message that this accuracy is achieved without knowing the details of the actual FTL.

To put these results into perspective, we repeated all the experiments with another state-of-the-art emulator called S.O.A [3]. For a fair comparison, since this tool embodies a fully reconfigurable FTL, its parameters were adjusted to provide the same WAF value used by SSDExplorer. This time the performance mismatch is 30% for sequential write and 70% for sequential read. When looking at random workloads, the S.O.A. emulator results are completely off. The roots of these discrepancies reside in the inability of the S.O.A emulator to accurately model the host interface command queuing, the multi-channel interleaving, and the ECC behavior. Therefore, S.O.A. emulators are not very well suited for FGDSE, which requires an accurate description of all the SSD components. It is interesting to point out that the mock-up SSD simulation results equal the results of a full drive simulation.

3.4.2 NVRAM Card

SSDExplorer was also tested on a 512 GB PCI-Express-NVRAM card for cloud applications [8]. These devices are used as a cache layer for file-systems' meta-data or for data intensive applications. In fact, as shown in Fig. 3.12, NVRAM cards incorporate a large DRAM buffer (usually between 8 and 16 GByte) which can be directly accessed by the user through the PCI-Express interface. Thanks to this feature, NVRAM cards can be used as a storage block device with a a sub-microsecond data access latency. Clearly, due to the volatile nature of DRAM memories, the card includes a super-capacitor for data protection during power-loss events. When the power drops, all data stored in the DRAM are flushed into the NAND flash memory array mounted on the same board. During this operation the NVRAM card behaves like a traditional multi-channel SSD: in this case, data stored in the DRAM play the role of the data usually coming from the host.

The simulated NVRAM card is configured with a single 8 GB DDR3 (1333 MT/s) DRAM buffer and a PCI-Express Gen2 x8 lanes host interface. The controller responsible for the data transfer from the on-board DRAM buffer to the persistent NAND flash array handles 8 channels, 4 targets each. NAND Flash memories are 2x-nm MLC devices with a page program time $t\_PROG = 1.8$ ms, a page read time $t\_READ = 115$ µs and a block erase time $t\_BERS = 6$ ms. Results for sequential write, sequential read, and random read workloads are shown in Fig. 3.13. Even with this very high performance architecture, SSDExplorer shows very low mismatch errors: about 0.01% for both read and write operations.

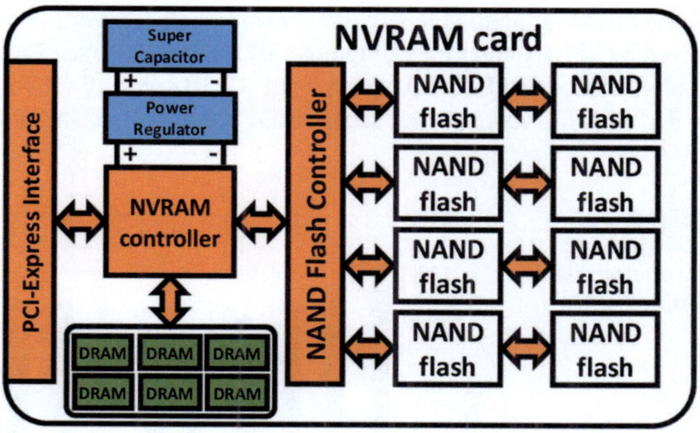

Fig. 3.12 NVRAM card architecture

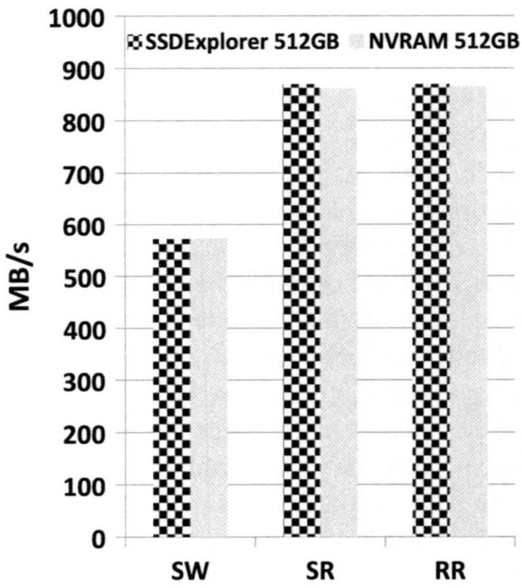

Fig. 3.13 Simulation accuracy of SSDExplorer (full disk simulation) in terms of throughput for Sequential Write (SW), Sequential Read (SR) and Random Read (RR). Device under test: 512 GB NVRAM card

3.5 Simulation Speed

SSDExplorer is completely written in SystemC and, therefore, its capability to be accurate pushes against simulation time. Since SSDExplorer includes PA-CA and TL-CA models, the number of *kilo-Cycles per Second* (kCPS) represents the only available metric for simulation speed, whereas the performance of emulation/simulation tools mainly based on behavioral models is measured in elapsed CPU time, thus making impossible any direct comparison. Figure 3.14 shows the kCPS achieved by SSDExplorer for 9 different SSD architectures (see Table. 3.3 for details) on an Intel Xeon CPU E5520 clocked at 2.27 GHz with 12 GByte of RAM and Linux Redhat. The workload is a sequential 4 kByte write distributed among

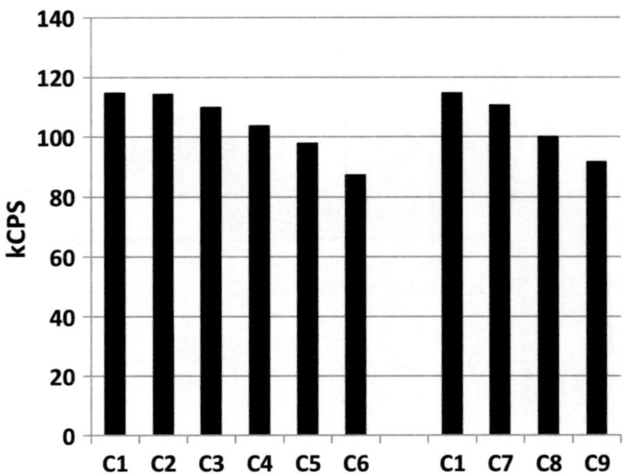

Fig. 3.14 SSDExplorer simulation speed with different SSD configurations using the WAF abstraction model

Table 3.3 SSD configurations used to evaluate SSDExplorer simulation speed

Configuration	SSD architecture
C1	1-CHN;1-TARGET
C2	2-CHN;1-TARGET
C3	4-CHN;1-TARGET
C4	8-CHN;1-TARGET
C5	16-CHN;1-TARGET
C6	32-CHN;1-TARGET
C7	1-CHN;2-TARGET
C8	1-CHN;4-TARGET
C9	1-CHN;8-TARGET

all the simulated NAND Flash targets, and the FTL is abstracted through the WAF model.

In Fig. 3.14 the first set of results (configuration C1–C6) shows the simulation speed dependency from the number of instantiated channels, while the second set of results (C1, C7–C9) shows the dependency from the number of NAND Flash targets. When the WAF model is replaced by a full FTL, simulation speed drops by a factor of three on average.

It is worth reporting that, even for resource-hungry configurations, the simulation speed is in the order of 100 kCPS which is definitely a great result for PC-CA EDA tools [43].

3.6 User Interface and WEB-Based Architecture

SSDExplorer is a complex simulator which requires a lot of inputs to enable a comprehensive FGDSE of the SSD design space. This situation is common to other simulation frameworks like DiskSim [5] and FlashSim [9], and it is a burden that users need to overcome if they want to have a fast and portable prototyping platform for SSD performance evaluation. To deal with these problems, SSDExplorer is provided with a complete Graphical User Interface (GUI) (sketched in Fig. 3.15) which guides the end user through all the parameters and the possible configurations of the tool.

The design of a specific SSD architecture is completely managed by the designers who can customize every single part of the SSD, from the memory architecture up to the host workload characteristics.

Scheduling a new simulation in SSDExplorer can be extremely simplified by using the available step-by-step configuration wizard. In essence, the user can build the desired SSD architecture by simply selecting among the blocks defined in the built-in database. In this way it is possible to either build a huge amount of different

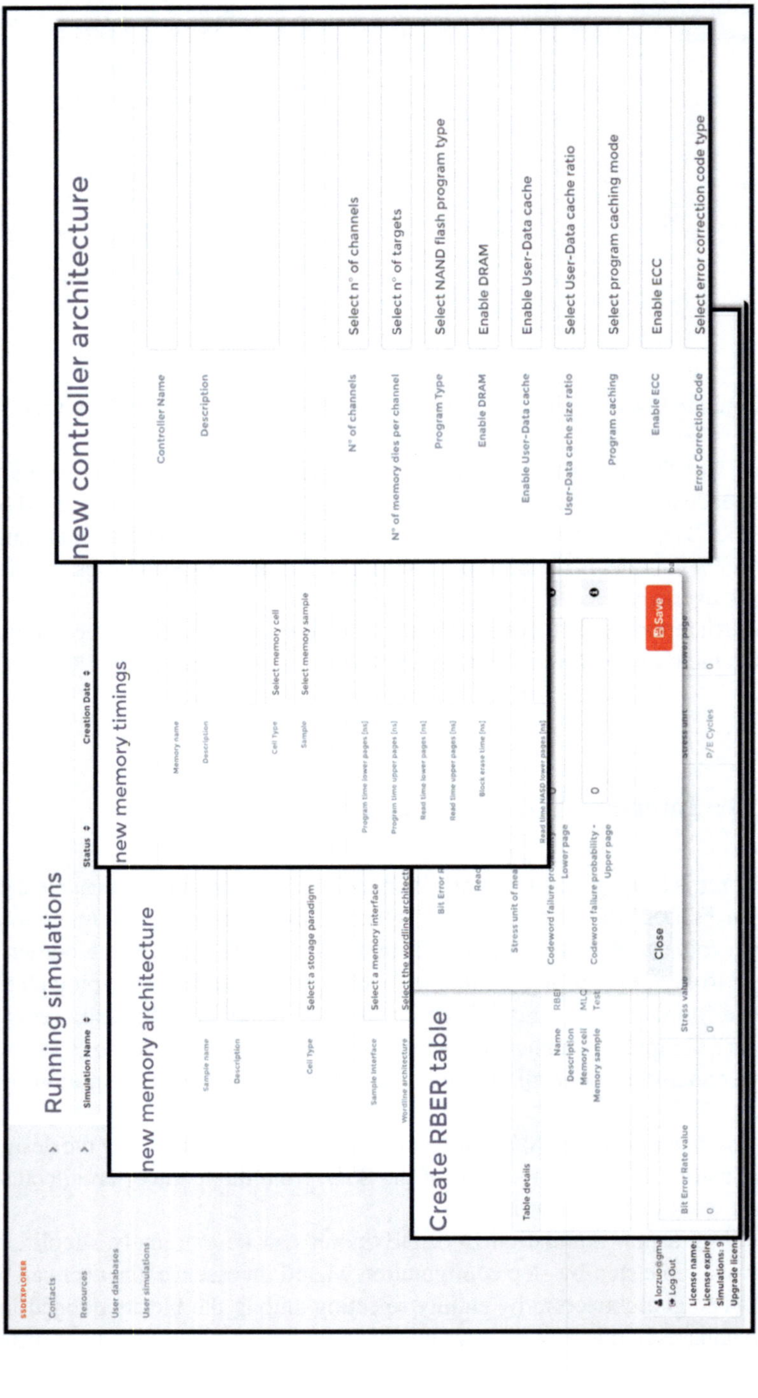

Fig. 3.15 SSDExplorer graphical user interface

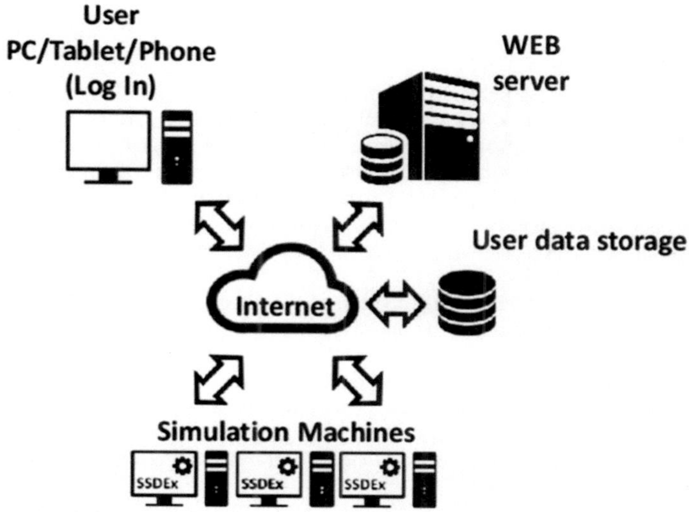

Fig. 3.16 SSDExplorer execution model

architectures or to modify a specific block without having to redefine the whole SSD architecture.

To further ease the usage of SSDExplorer, simulation results are available either in a text or in a graph format. In this way, if a deep data analysis or a further step of post-processing is required, it is possible to download the data in a portable Comma-Separated-Values (CSV) format.

Finally, to solve the common software portability and compatibility issues (i.e. the possibility of executing the software on all possible PC configurations), SSDExplorer adopts a cloud-based approach. It is worth mentioning that it is the first time that a distributed cloud-based execution model is adopted for simulating and optimizing an SSD. As sketched in Fig. 3.16, the whole source code is executed on an array of remote secure servers which exposes only a minimal set of APIs. These APIs are called by a WEB server which collects all the configuration parameters and the user's data on an encrypted database. Thanks to this approach, SSDExplorer can be easily used and ported on either a PC, a Tablet, or even a Smartphone. The only requirement is to have a WEB browser. Moreover, the cloud based approach gives the possibility of dynamically changing the number of SSDExplorer HW instances when needed, thus enabling a high level of scalability, which is a key requirement for every CAD tool.

The SSDExplorer simulation framework is available online at: https://ssdexplorer.azurewebsites.net.

References

1. R. Micheloni, A. Marelli, and K. Eshghi. *Inside Solid State Drives (SSDs)*. Springer, 2012.
2. The OpenSSD Project. http://www.openssd-project.org/wiki/The_OpenSSD_Project.
3. Jinsoo Yoo, Youjip Won, Joongwoo Hwang, Sooyong Kang, Jongmoo Choi, Sungroh Yoon, and Jaehyuk Cha. Vssim: Virtual machine based ssd simulator. In *IEEE Symposium on Mass Storage Systems and Technologies (MSST)*, pages 1–14, 2013.
4. QEMU: open source processor emulator. http://wiki.qemu.org/Main_Page.
5. The DiskSim simulation environment version 4.0, 2008. http://www.pdl.cmu.edu/PDL-FTP/DriveChar/CMU-PDL-08-101.pdf.
6. Xiao-Yu Hu, Evangelos Eleftheriou, Robert Haas, Ilias Iliadis, and Roman Pletka. Write amplification analysis in flash-based solid state drives. In *Proceedings of SYSTOR*, pages 10:1–10:9, 2009.
7. Ocz vertex series 120GB SSD. http://ocz.com/consumer.
8. FLASHTEC NVRAM Drives. http://pmcs.com/products/storage/flashtec_nvram_drives/.
9. Youngjae Kim, B. Tauras, A. Gupta, and B. Urgaonkar. Flashsim: A simulator for nand flash-based solid-state drives. In *International Conference on Advances in System Simulation (SIMUL)*, pages 125–131, 2009.
10. Sungjin Lee, Kermin Fleming, Jihoon Park, Keonsoo Ha, Adrian M. Caulfield, Steven Swanson, Arvind, and Jihong Kim. Bluessd: An open platform for cross-layer experiments for nand flash-based ssds. In *The 5th Workshop on Architectural Research Prototyping*, 2010.
11. Jongmin Lee, Eujoon Byun, Hanmook Park, Jongmoo Choi, Donghee Lee, and Sam H. Noh. Cps-sim: configurable and accurate clock precision solid state drive simulator. In *Proceedings of the ACM symposium on Applied Computing*, pages 318–325, 2009.
12. Hoeseung Jung, Sanghyuk Jung, and Yong Ho Song. Architecture exploration of flash memory storage controller through a cycle accurate profiling. *IEEE Transactions on Consumer Electronics*, 57(4):1756–1764, 2011.
13. E.-Y. Chung. A Solid-State Disk Simulator for Quantitative Performance Analysis and Optimization. In *NVRAMOS*, 2009.
14. Cagdas Dirik and Bruce Jacob. The performance of pc solid-state disks (ssds) as a function of bandwidth, concurrency, device architecture, and system organization. In *International Symposium on Computer Architecture (ISCA)*, pages 279–289, 2009.
15. S. Zertal and W. Dron. Quantitative study of solid state disks for mass storage. In *International Symposium on Performance Evaluation of Computer and Telecommunication Systems (SPECTS)*, pages 149–155, 2010.
16. Kai Zhao, Wenzhe Zhao, Hongbin Sun, Xiaodong Zhang, Nanning Zheng, and Tong Zhang. Ldpc-in-ssd: Making advanced error correction codes work effectively in solid state drives. In *Presented as part of the 11th USENIX Conference on File and Storage Technologies (FAST 13)*, pages 243–256, 2013.
17. Systemc 2.0.1 language reference manual, 2002. http://www.systemc.org.
18. Sungpack Hong, Sungjoo Yoo, Sheayun Lee, Sangwoo Lee, Hye Jeong Nam, Bum-Seok Yoo, Jaehyung Hwang, Donghyun Song, Janghwan Kim, Jeongeun Kim, HoonSang Jin, Kyu-Myung Choi, Jeong-Taek Kong, and SooKwan Eo. Creation and utilization of a virtual platform for embedded software optimization:: an industrial case study. In *Proceedings of the International Conference Hardware/Software Codesign and System Synthesis (CODES+ISSS)*, pages 235–240, Oct 2006.
19. S. Pasrich and N. Dutt. *On-Chip Communication Architectures: System on Chip Interconnect*. Morgan Kaufmann, 2008.
20. Nathan Binkert, Bradford Beckmann, Gabriel Black, Steven K. Reinhardt, Ali Saidi, Arkaprava Basu, Joel Hestness, Derek R. Hower, Tushar Krishna, Somayeh Sardashti, Rathijit Sen, Korey Sewell, Muhammad Shoaib, Nilay Vaish, Mark D. Hill, and David A. Wood. The gem5 simulator. *SIGARCH Comput. Archit. News*, 39(2):1–7, 2011.
21. Open Virtual Platforms - the source of Fast Processor Models and PlatformsOpen Virtual Platforms. http://www.ovpworld.org.

22. Open Nand Flash Interface (ONFI). http://www.onfi.org.
23. Nand flash interface interoperability, 2014. www.onfi.org/~/media/onfi/specs/jesd230b.pdf?la=en.
24. Evatronix NAND Flash controller ip-core. http://www.evatronix-ip.com/ip-cores/memory-controllers/nand-flash.html.
25. Serial ATA International Organization. *SATA revision 3.0 specifications*. www.sata-io.org.
26. SATA-IP host reference design on SP605 manual, Apr 2013. Accessed.
27. NVM Express, 2013. http://www.nvmexpress.org/.
28. Nvm express 1.1 specification, 2013. http://nvmexpress.org/wp-content/uploads/2013/05/NVM_Express_1_1.pdf.
29. Open-Silicon. SATA device controller - product brief, 2013. http://www.open-silicon.com/ip-technology/open-silicon-ip/io-controllers/sata-device-controller/.
30. Mt41j256m8 datasheet - micron technology, 2006. chrome-extension://encfpfilknmenlmjemepncnlbbjlabkc/https://www.micron.com/ /media/documents/products/data-sheet/dram/ddr3/2gb_ddr3_sdram.pdf.
31. Intel X18-M X25-M SATA Solid State Drive. Enterprise Server/Storage Applications. http://cache-www.intel.com/cd/00/00/42/52/425265_425265.pdf.
32. P. Rosenfeld, E. Cooper-Balis, and B. Jacob. Dramsim2: A cycle accurate memory system simulator. *IEEE Computer Architecture Letters*, 10(1):16–19, 2011.
33. Myoungsoo Jung, E.H. Wilson, D. Donofrio, J. Shalf, and M.T. Kandemir. Nandflashsim: Intrinsic latency variation aware nand flash memory system modeling and simulation at microarchitecture level. In *IEEE 28th Symposium on Mass Storage Systems and Technologies (MSST)*, pages 1–12, 2012.
34. Youngjoo Lee, Hoyoung Yoo, Injae Yoo, and In-Cheol Park. 6.4gb/s multi-threaded bch encoder and decoder for multi-channel ssd controllers. In *IEEE International Solid-State Circuits Conference Digest of Technical Papers (ISSCC)*, pages 426–428, Feb 2012.
35. L. Zuolo, C. Zambelli, R. Micheloni, D. Bertozzi, and P. Olivo. Analysis of reliability/performance trade-off in solid state drives. In *IEEE International Reliability Physics Symposium*, pages 4B.3.1–4B.3.5, June 2014.
36. Duo Liu, Yi Wang, Zhiwei Qin, Zili Shao, and Yong Guan. A space reuse strategy for flash translation layers in slc nand flash memory storage systems. *Very Large Scale Integration (VLSI) Systems, IEEE Transactions on*, 20(6):1094–1107, June 2012.
37. Tianzheng Wang, Duo Liu, Yi Wang, and Zili Shao. Ftl2: A hybrid flash translation layer with logging for write reduction in flash memory. *SIGPLAN Not.*, 48(5):91–100, June 2013.
38. Yuan-Hao Chang, Po-Chun Huang, Pei-Han Hsu, L.-J. Lee, Tei-Wei Kuo, and D.H.-C. Du. Reliability enhancement of flash-memory storage systems: An efficient version-based design. *Computers, IEEE Transactions on*, 62(12):2503–2515, Dec 2013.
39. Intel Shows PAX Attendees SSD Overclocking. http://www.legitreviews.com/intel-shows-pax-attendees-ssd-overclocking_122557.
40. Indilix barefoot controller. http://www.indilinx.com/solutions/barefoot.html.
41. IOzone Filesystem Benchmark. http://www.iozone.org/.
42. UMassTraceRepository. http://traces.cs.umass.edu/index.php/Storage/Storage.
43. L. Benini, D. Bertozzi, A. Bogliolo, F. Menichelli, and M.Olivieri. Mparm: Exploring the multi-processor soc design space with systemc. *Journal of VLSI SIgnal Processing*, 41:169–182, 2005.

Chapter 4
Design Trade-Offs for NAND Flash-Based SSDs

Lorenzo Zuolo, Cristian Zambelli, Alessia Marelli, Rino Micheloni and Piero Olivo

Abstract During the design phase of an SSD, the target application must be always taken into account. In fact, depending on the most important requirements, such as bandwidth, latency or reliability, architecture and cost of the drive might be totally different. Therefore, at the beginning of the development process, a thorough design space exploration is strongly recommended. In this chapter we describe the main actors of the drive architecture and we show how a dedicated CAD tool such as SSDExplorer can be used to optimize the SSD design, given a set of constraints. In particular, we consider 3 different cases: design for maximum bandwidth, design for minimum latency, and performance/reliability trade-off. Of course, in all cases SSD simulations are used to identify the right architecture to achieve the design target, while minimizing the resource request (e.g. number of Flash channels, number of NAND Flash memories, number of processor cores, etc...). For each case, this chapter includes design examples and corresponding simulation results.

During the design phase of an SSD, the target application must be always taken into account. In fact, depending on the most important requirements, such as bandwidth, latency or reliability, architecture and cost of the drive might be totally different. Therefore, at the beginning of the development process, a thorough design space exploration is strongly recommended. In this chapter we describe the main actors of the drive architecture, when looking at bandwidth and latency.

L. Zuolo (✉) · C. Zambelli · P. Olivo
Dipartimento di Ingegneria, Università degli Studi di Ferrara, via G. Saragat, 1, 44122 Ferrara, Italy
e-mail: lorenzo.zuolo@unife.it

A. Marelli
Microsemi, Via Torri Bianche 1, 20871 Vimercate, Italy

R. Micheloni
Performance Storage Business Unit, Microsemi Corporation, Vimercate, Italy

R. Micheloni (ed.), *Solid-State-Drives (SSDs) Modeling*,
Springer Series in Advanced Microelectronics 58,
DOI 10.1007/978-3-319-51735-3_4

4.1 Design for Maximum Performance

In this Section we show how a dedicated CAD tool such as SSDExplorer can be used to optimize the SSD design, given a target performance. The goal is to minimize resources while achieving the target host interface bandwidth. Table 4.1 shows a set of representative design points which were used for this investigation. All simulations have been performed with a synthetic workload made of 4 kB sequential writes. All data were collected by using two well known DRAM buffer management policies [1]: *write back and write through caching* and *no caching* [2]. With the former approach, the SSD controller acknowledges the end of each transaction to the host, i.e. when data are moved from the host interface to the DRAM buffers. In the latter case, the acknowledgment happens only when all data have been actually written in the NAND flash memory. All experimental results are based on a 4x-nm MLC NAND, whose main characteristics are t_PROG from 900 μs to 3 ms, $t\_READ = 60$ μs and t_BERS from 1 ms to 10 ms [3].

Figure 4.1 shows how the different architectures perform with a SATA II host interface. The *SATA ideal* curve represents the limit of the interface itself. The *SATA+DDR* curve is more realistic since it incorporates the time spent by its internal DMA engines to transfer the data from the host system to the DRAM buffers (i.e. the time required to process the transactions from the host). The best design option is the one that gets closer to the *SATA+DDR* bandwidth by maximizing the bandwidth of the *DDR+FLASH* curve (i.e. the time spent by the flash memory to flush the DRAM buffer and write the data). The *SSD cache/ SSD no cache* curves represent the bandwidth of the entire drive.

When caching is used, *C6*, *C8* and *C10* are the best solutions since they reach the target performance and saturate the host interface bandwidth. However, if we take into account the hardware costs, *C6* represents the right choice since it requires the lowest amount of resources. On the other hand, when *no caching* is used, the overall

Table 4.1 SSD configurations for design exploration. DDR-buf = DDR Buffer, CHN = Flash Channel

Configuration	SSD architecture
C1	4-DDR-buf;4-CHN;4-TARGET;2-DIE
C2	8-DDR-buf;8-CHN;4-TARGET;2-DIE
C3	8-DDR-buf;8-CHN;8-TARGET;2-DIE
C4	8-DDR-buf;8-CHN;8-TARGET;4-DIE
C5	8-DDR-buf;8-CHN;8-TARGET;8-DIE
C6	16-DDR-buf;16-CHN;8-TARGET;4-DIE
C7	16-DDR-buf;16-CHN;4-TARGET;2-DIE
C8	32-DDR-buf;32-CHN;4-TARGET;2-DIE
C9	32-DDR-buf;32-CHN;1-TARGET;1-DIE
C10	32-DDR-buf;32-CHN;8-TARGET;4-DIE

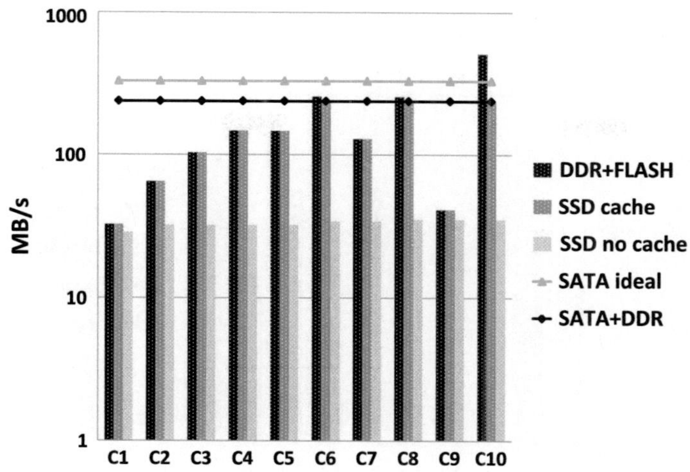

Fig. 4.1 Comparison of the configurations proposed in Table 4.1. Sequential write with SATA II host interface

drive performance (*SSD no cache*) is strongly limited, and none of the configurations can meet the target.

The reason behind the performance flattening with *no caching* lies on the SATA interface and, in particular, into its limited command queue depth. In fact, the SATA protocol is able to manage up to a maximum of 32 commands at once. Therefore, with a *no caching* policy, the host interface cannot accept new commands until the queue is cleared (i.e. commands are committed to the NAND memories). As a result, the internal parallelism can't be leveraged. Indeed, the *DDR+FLASH* bar clearly indicates that the bandwidth could be much higher.

Figure 4.2 shows the simulation results with a PCI-Express Gen2 interface featuring 8 lanes and the NVM-Express protocol. Due to the higher speed, the host interface does no longer represent the performance bottleneck. In fact, even the configuration with the highest parallelism (*C10*) is not able to saturate the interface. The major result of Fig. 4.2 becomes clear when looking at the *SSD no caching* bars. In this case, since the NVM-Express protocol can handle up to 64k-commands, the SSD internal parallelism can be fully exploited. A performance gap between the different configurations still exists. Indeed, the time spent to flush the incoming data to the NAND flash memories can be hidden with cached architectures. If maximum performance is the main driver, *C10* is the best solution; otherwise, if the performance-cost trade-off matters the most, solutions ranging from *C3* to *C8* are all eligible.

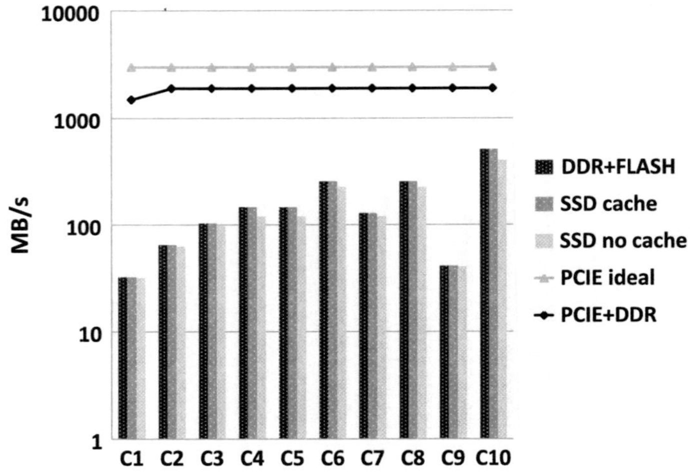

Fig. 4.2 Sequential write with PCIe host interface. Comparison of the configurations proposed in Table 4.1

4.2 Design for Minimum Latency

As described in Sect. 4.1, the main parameters impacting the bandwidth are the internal controller parallelism and the host interface *Queue Depth* (QD). This latter, however, has a severe impact on another important SSD metric: the *Quality of Service* (QoS) [4]. In an SSD, the QoS is calculated at a percentile (e.g. 99.99) of the SSD's latency distribution and it is used to quantify how the SSD behaves in the worst-case conditions. By using this metric we can understand if the target drive architecture is suitable for a specific application, such as real-time and safety-critical systems. Figure 4.3 shows how bandwidth and QoS scale when the host QD ranges from 1 to 256 commands. Simulations are based on 8 channels with 8 mid-1x nm TLC NAND flash targets. Average read time is 86 μs and workload is 100% 4 kB random read. Takeaway from Fig. 4.3 is: the bigger the host interface queue depth is, the higher the SSD QoS is. This behavior, however, is in contrast with the requirements of high performance SSDs, which ask for achieving the target bandwidth with the shortest QoS. In fact, nowadays user applications such as financial transactions or e-commerce platforms [5] are designed to work with storage devices which have to serve an I/O operation within a specific time-frame which is usually upper-bounded by the QoS requirement.

In order to deal with this performance/latency trade-off, usually the host interface is set to work with a specific QD. In such a way the design space is reduced and the optimization becomes easier. For instance, if a 300 kIOPS bandwidth and a 2 ms QoS are required (solid lines in Fig. 4.3), the configurations shown in Fig. 4.3 indicate that a QD of 64 commands is the preferred choice. One of the main limitations of the design methodology presented so far is the lack of QD flexibility, which is usually required

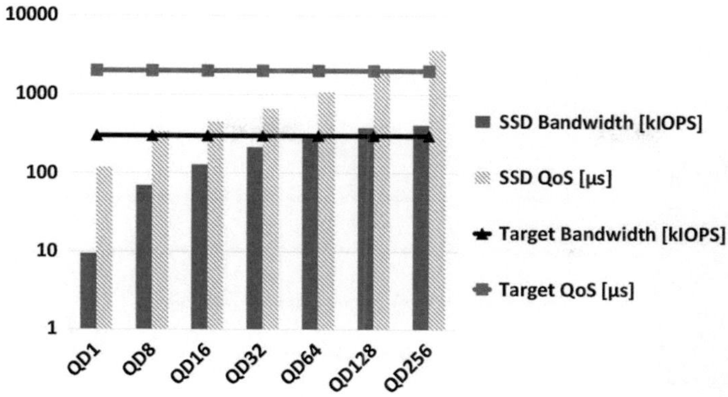

Fig. 4.3 SSD bandwidth and QoS as a function of the host *Queue Depth*

by the host system. In fact, during the design phase of an SSD, it is common to define only the target bandwidth, because it is not known a priori on which application the drive will be used. For example, SSDs used in flexible I/O environments, where users can instantiate complete virtual platforms in a "plug-and-play" fashion [6], have to guarantee the target performance and QoS for a general purpose traffic. As a consequence, since in these systems the actual host QD is strictly dependent upon the number of parallel threads on which the target application is running, the number of commands that will be issued by the user and the corresponding QD cannot be predicted. Therefore, it is clear that, to meet the general purpose workload specifications of these environments, the SSD has to be designed to provide the target performance and QoS independently from the host QD.

To deal with this requirement it is possible to use the *Head-of-Line* (HoL) blocking, whose effect is to limit the number of outstanding commands inside the SSD. To clarify this point, it is useful to study how host commands are managed and queued by the SSD controller.

Figure 4.4 shows the queuing hierarchy usually implemented in traditional SSD controllers. Besides the external host QD, it is common to have a dedicated small command queue for each NAND flash memory die: the *Target Command Queue* (TCQ). Basically, thanks to the TCQ, the host can continue to issue commands even when it tries to read or program a chip which is already in the busy state. In fact, when this condition is verified, the operation is simply queued in the TCQ and the SSD controller can continue to fetch other commands from the host QD. This approach, on one hand allows maximizing the bandwidth since TCQs keep always busy all the NAND flash memories; on the other hand, however, it increases the number of outstanding commands inside the SSD, because several operations have to wait inside the TCQs before being served. At this point it is clear that, when there are a lot of commands inside the drive (i.e. when a large host QD is used), the SSD controller and all the related TCQs fall in a deep saturation state which leads to an increased QoS.

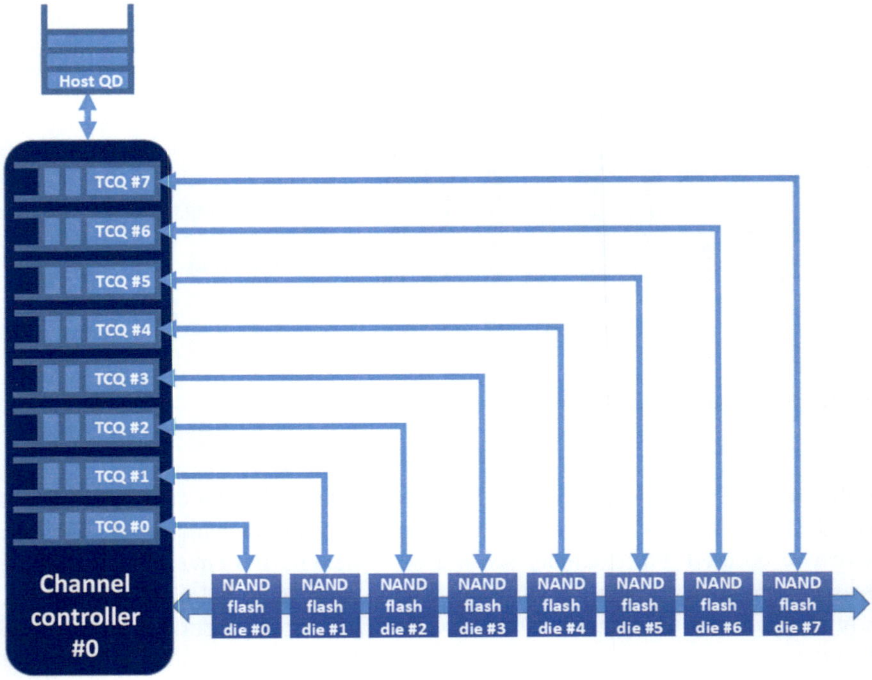

Fig. 4.4 Queueing hierarchy implemented inside the SSD controller

When the HoL blocking is used, the aforementioned latency problem can be partially solved. When the number of commands queued in single TCQ exceeds a predefined threshold, it is possible to trigger a blocking state inside the SSD controller which stops the submission of a new command from the host QD. In such a way, depending on the HoL threshold value, it is possible to avoid long command queues inside the TCQs and, hence, the disk's QoS can be limited within a specific window.

Figure 4.5 shows the effectiveness of the HoL blocking. Simulations are based on the same SSD configurations and workload of the previous section. Unlike the case presented in Fig. 4.3, as soon as the target performance of 300 kIOPS is reached (QD64 configuration), the HoL blocking effect starts keeping the QoS below the target requirements even when long QDs, such as QD128 and QD256, are considered.

The fine-grained QoS calibration made available by the HoL blocking, however, does not come for free. With reference to Fig. 4.6, if, besides the bandwidth and QoS, the average SSD latency is taken into account, it is clear that the HoL Blocking effect has to be wisely used. As a matter of fact, when the HoL blocking is triggered, it trades the QoS reduction with an increase of the average latency. Moreover, this behavior becomes more pronounced when high QDs are used (QD64, QD128, and QD256), i.e. when a higher QoS reduction is required.

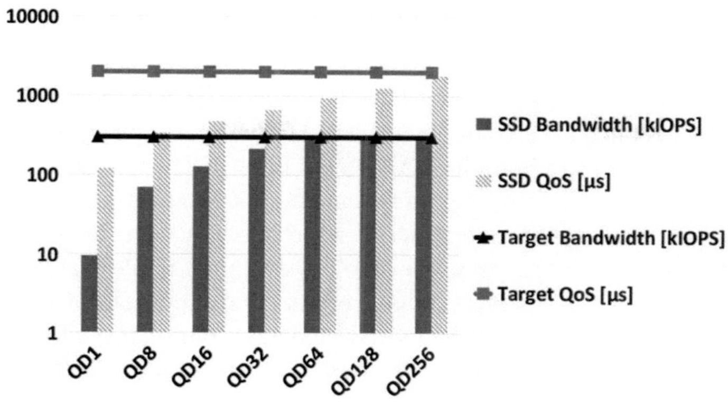

Fig. 4.5 SSD bandwidth and QoS as a function of the host queue depth (QD) when HoL blocking is used

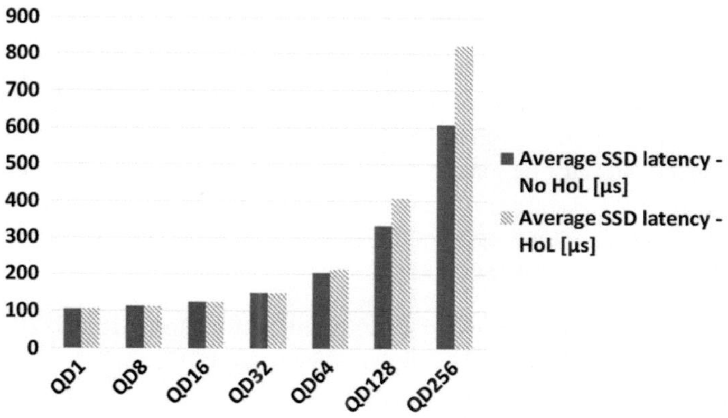

Fig. 4.6 Average SSD latency evaluated with and without the HoL blocking

4.3 Performance/Reliability Trade-Off

Achieving high performances (i.e. high bandwidth and low latency) with the lowest possible resource allocation is one of the main goals of the most modern SSD architectures. However, even if high read/write bandwidth can be easily achieved, one of the main problem of SSDs is the data reliability, which is dependent on the non-volatile NAND flash memories used as the storage medium. These components are subject to a progressive wear-out whose physical roots reside in the tunnel oxide degradation related to the Fowler-Nordheim tunneling adopted during program/erase. Such mechanism leads to a variation of the Floating Gate charge which translates into a threshold voltage shift [7]. A direct indication of this phenomenon is a progressive increase of the *Raw Bit Error Rate* (RBER) of the memory. The RBER is the percentage of bits in error after a single read operation [8]. Such an increase translates into the inability of retrieving the correct data either after a number of Program/Erase operations (i.e., P/E cycles) or after a long retention time.

Error Correction Codes (ECC) such as BCH and LDPC are required to improve the reliability of stored data, while sophisticated write algorithms control the threshold voltage distributions inside the NAND chip [9, 10]. The relentless demand for storage capacity has forced the migration from *Single-Level Cell* (SLC) to more complex *Multi-Level Cells* (MLC) memories, in which more than one bit is stored within a single physical cell. This aggressive technology evolution, while enabling the usage of SSDs in big-data cloud servers, it also results in an exponential RBER increase [11]. To deal with increasing RBERs, NAND and SSD controller vendors have introduced new read techniques, such as *Read Retry* (RR) and LDPC *Soft Decoding* reads (SD) [12]. Basically, these approaches are based on repeating the read operation of the page in error to either reduce the RBER or to exploit the strong soft decoding capabilities of the ECC (ECC_{th}) [13].

These procedures, however, inevitably imply an overall performance degradation since a single page is available only after several read operations [14]. Moreover, although these algorithms can enhance the NAND reliability, the memory controller has to be modified to carefully handle them. The straightforward conclusion is that the performance/reliability trade-off introduced by the emerging algorithms for reliability enhancement must be mapped to the performance/reliability trade-off of the entire SSD.

In the next Sections a thorough evaluation of the performance/reliability trade-off in SSDs will be performed showing that:

- RR and SD algorithms were developed to significantly improve the SSD reliability, without considering the impact on the performances of the drive;
- RR and SD introduce a read bandwidth degradation which is difficult to mask. In fact, while several disk operations such as wear leveling, garbage collection and bad block management can be masked at the firmware level (for instance by background execution) [9, 10], both the SD and the RR algorithms cannot be masked by the firmware, since they have to directly access the storage medium. Therefore, only architectural countermeasures, such as high performance host system interface and a stronger ECC (to reduce RR and SD trigger rate), seem to be viable solutions.

4.3.1 Read Retry

RR are specific algorithms embedded inside the NAND flash die, which can significantly reduce the RBER of the memory during both endurance (i.e. program/erase cycles) and retention. These approaches leverage a fine tuning of the internal NAND read voltage, which is adjusted depending on the actual wear-out state of the device [15, 16]. Thanks to this shift of the read reference voltages, it is possible to limit the

Fig. 4.7 Example of RBER in a commercial 2x-nm node MLC NAND flash as a function of the number program/erase (P/E) cycles. Average values for upper pages and lower pages, and for the entire memory are shown

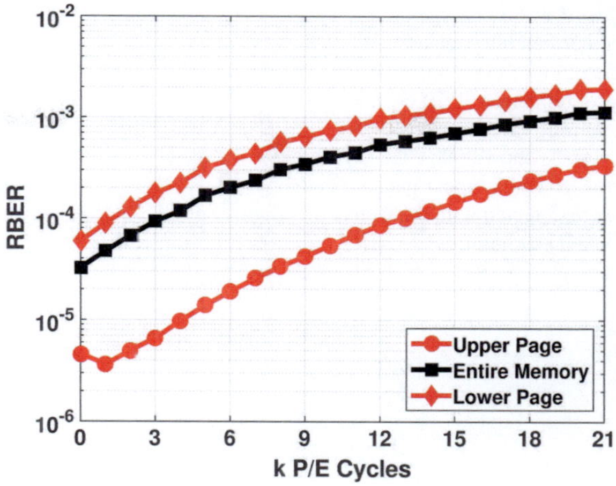

Fig. 4.8 Example of RBER in a commercial 2x-nm node MLC NAND flash as a function of retention time after a pre-defined number of P/E cycle. Average values for upper pages and lower pages, and for the entire memory are shown

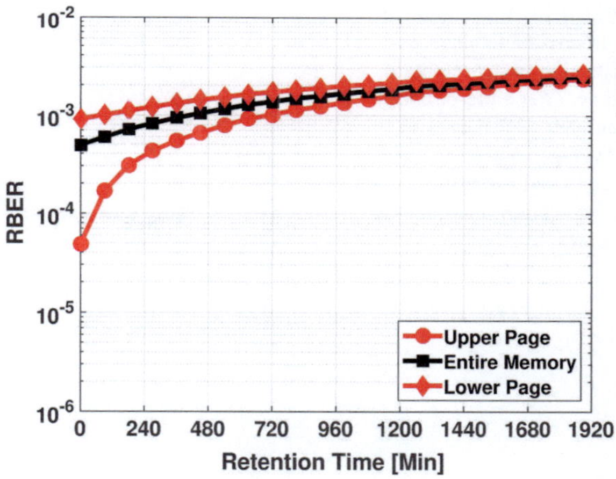

number of read errors and, hence, to reduce the ECC correction efforts in terms of both power consumption and time.

Figures 4.7 and 4.8 show the RBER for a MLC NAND flash memory manufactured in a 2x-nm node as a function of the number of P/E cycles and of the retention time at a specific temperature after a pre-defined number of P/E cycles, respectively. Data have been collected up to twice the rated endurance of the memory. These figures show the different average raw BER values for upper and lower pages, and for the entire memory, respectively.

A higher RBER translates into a higher probability of erroneously decoding the bits read from the NAND (i.e. uncorrectable pages). As a consequence, when the number of erroneous bits in a page exceeds the error correction capability of the code ECC_{th}, the page content can't be retrieved anymore. Figures 4.9 and 4.10 show the percentage of read failures for endurance and retention, respectively, when no RBER

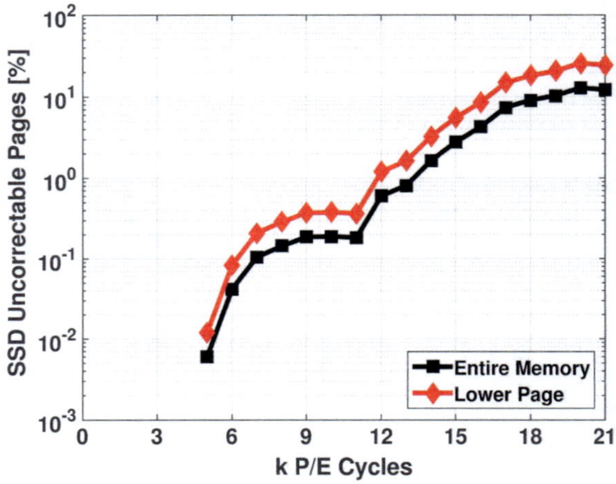

Fig. 4.9 Percentage of uncorrectable pages as a function of the number of program/erase (P/E) cycles. Average value for lower pages is shown. No uncorrectable upper pages were detected with the tested NAND technology

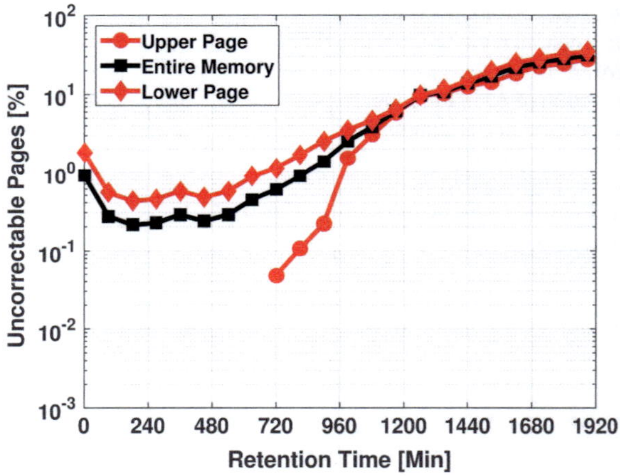

Fig. 4.10 Percentage of uncorrectable pages as a function of the retention time, after a pre-defined number of P/E cycles. Average values for upper pages and lower pages, and for the entire memory are shown

reduction techniques are applied. In this case the ECC is able to correct up to 100 bits in error in a 4320 Bytes codeword (ECC_{th}). The different behavior of endurance and retention is due to the different drifts of the threshold voltage distributions.

Figures 4.11 and 4.12 show the effectiveness of RR in reducing the NAND RBER. Both endurance and retention benefit from Read Retry.

However, even if the RR algorithm is able to reduce the RBER of a NAND flash memory, when it is used inside an SSD it has to be wisely called. Indeed, whenever the ECC detects that a page cannot be corrected, it can request for a RR intervention. In this case, in order to reduce the RBER, the page is read again from the memory with a set of modified reference voltages, which delay the access to the memory. Moreover, this process could be required several times, i.e. until the ECC can correct the addressed page. Since the occurrence of an uncorrectable page event strongly depends upon the Flash technology characteristics and its actual usage model, the

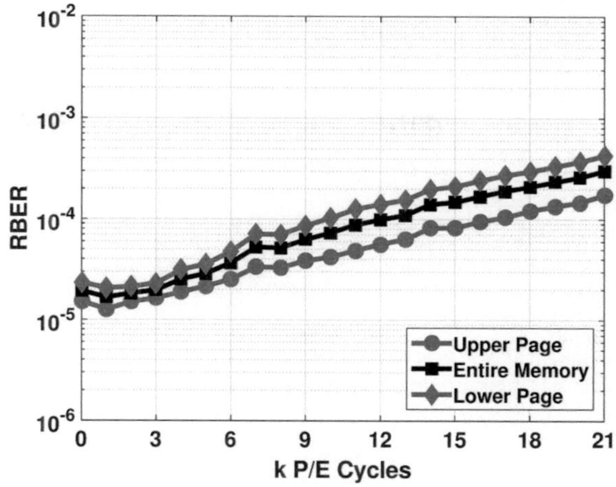

Fig. 4.11 NAND RBER after RR intervention as a function of the number of program/erase (P/E) cycles. Average values for upper pages and lower pages, and for the entire memory are shown

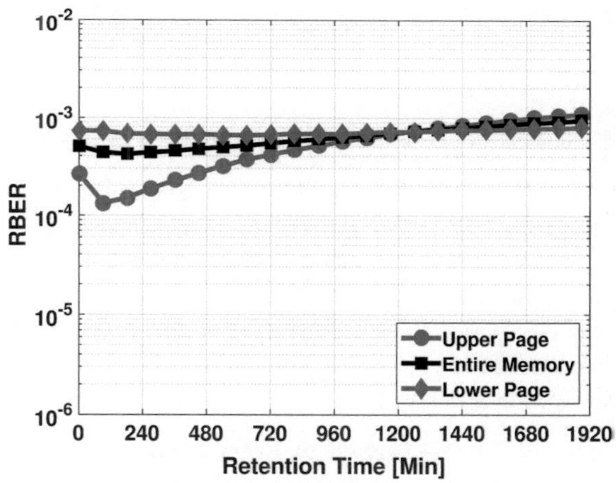

Fig. 4.12 NAND RBER after RR intervention as a function of the retention time after a pre-defined number of P/E cycles. The average values for the upper and lower pages and that of the entire memory are shown

RR intervention cannot be easily predicted; as such, a longer read latency might reduce the SSD bandwidth below acceptable values, unless RR is properly taken into account at the architectural level.

To test how RR techniques impact SSD's performances, 4 different drive architectures have been simulated, as summarized in Table 4.2. It is worth pointing out that, even if the SSD's architecture was modified from one simulation to another, the drive capacity was kept constant, to avoid any possible inconsistency in analyzing the results. As stated in Sect. 4.3.1, the two main components that contribute to the RR-induced drawbacks are the NAND flash itself and the ECC. In order to accurately reproduce their behavior, the ECC engine has been modeled assuming a BCH code able to correct up to 100 errors in a 4320 bytes codeword. For the latency modeling of the BCH decoder, which is the ECC block active during a read operation, the internal

Table 4.2 SSD's architectures for read retry analysis

Parameter	Architecture			
Host interface	SATA III			
DDR-buffer	1			
ECC	1x Channel			
Disk capacity	1 TB			
Configurations	C1	C2	C3	C4
Channels	1	2	4	8
NAND flash targets per channel	16	8	4	2

syndrome and Berlekamp-Massey modules follow the state of the art implementation provided in [17], whereas the Chien search machine is modeled as follows:

- a faster (i.e. higher parallelism) hardware to recover a relative small number of errors;
- a slower hardware to correct up to the maximum number of allowed errors.

The NAND flash adopted for these simulations is a 2x-nm MLC technology with a page size of 16 kbyte (plus Spare bytes) and a block size of 8 Mbyte. RBER values, access times and the uncorrectable page event frequency have been characterized through lab measurements and back-annotated into the SSD simulator. Starting from the RBER value, the corresponding number of errors is extracted from the probability density function [18], and then fed into the modeled ECC engine in order to extract the decoding delay.

Section 4.3.1 **(A) Impact of Retry impact on SSD'a performances**. Figure 4.13 shows the simulated architecture which includes a single BCH engine connected to a single channel serving 16 memory targets, each made of two dies.

For the sake of clarity and to avoid any artifact in the performance analysis due to the host interface model (e.g. SATAIII [19] or PCI-Express NVM-Express [20, 21]), this last was neglected. This approach allows focusing on the actual bandwidth implications of Flash memories and ECC.

The main outcome of this analysis is that the read bandwidth tracks the memory wear-out, as displayed in Fig. 4.14. Since the real SSD behavior depends on P/E cycles and retention time, the adoption of RBER as a reference metric allows neglecting the actual degradation mechanism. Two points are highlighted in the figure: a 2% drop (point A) and a 10% drop (point B) with respect to the SSD read bandwidth at *Beginning of Life* (BoL), respectively. The former point will be used to show in details the behaviors of the ECC engine and of the NAND memories in a "normal" read condition whereas the latter point, which represents the maximum acceptable degradation level, will be used to illustrate architectural limitations when the number of uncorrectable pages becomes the dominant factor.

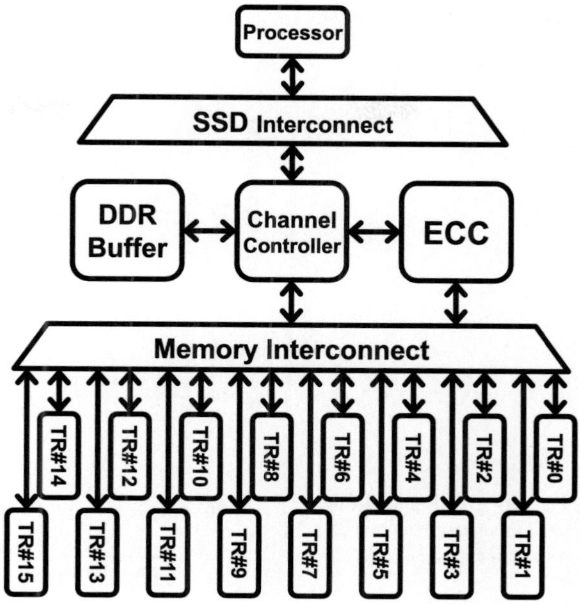

Fig. 4.13 SSD's reference architecture used in this section

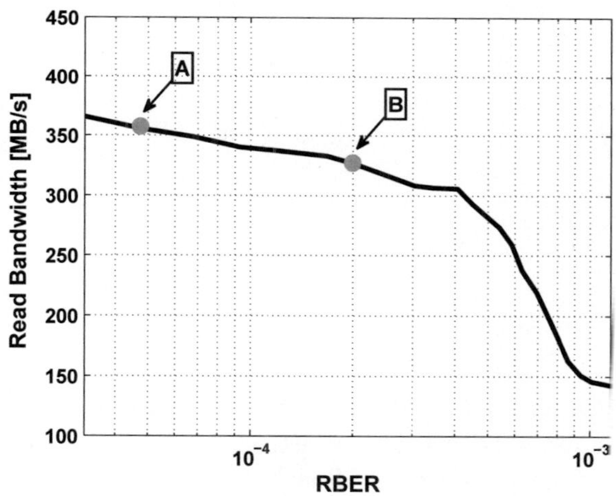

Fig. 4.14 SSD read bandwidth versus RBER for the reference architecture of Fig. 4.13. Points A and B represent a 2% and a 10% performance degradation compared to the BoL, respectively

Let's start from point A. Figure 4.15 shows the distribution of the correction times spent by the ECC engine, which, of course, depends on the number of errors to be corrected. Latency distribution is shown in Fig. 4.16, where the read scheduling mechanism is highlighted. The ECC correction time is one order of magnitude shorter than the memory read latencies and, therefore, it has a marginal impact on the performance of the drive.

Figures 4.17 and 4.18 show what happens at point B. In particular, a broadening in the ECC correction time distribution can be observed. The maximum correction time,

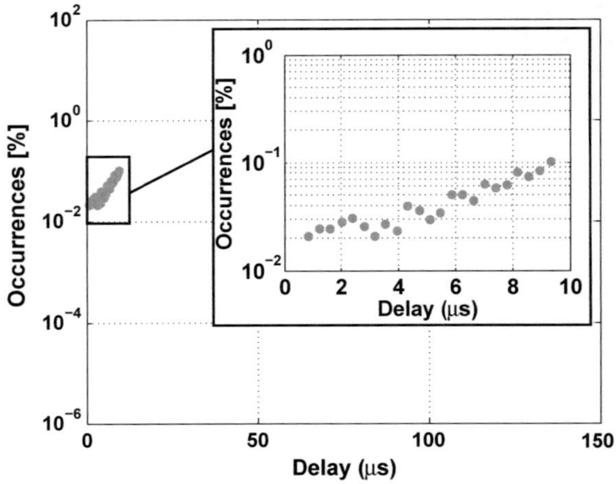

Fig. 4.15 Distribution of the ECC correction times at point A of Fig. 4.14

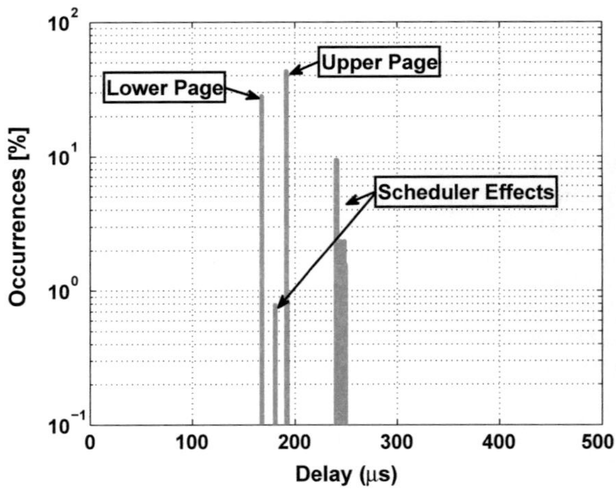

Fig. 4.16 Distribution of the NAND memory latencies at point A of Fig. 4.14

identified in Fig. 4.17, is related to the case of an uncorrectable page. Figure 4.18, on the other hand, shows that, when there is a significant number of uncorrectable pages, the consequent RR activation adds two components to the NAND flash read latency: the first one is related to the commands used to activate RR, while the second one is dictated by the actual execution time of the RR algorithm.

Finally, it is worth pointing out that the percentage of uncorrectable page reads that produces a 10% drop in the SSD read bandwidth is in the order of 0.6–0.8%.

Section 4.3.1 **(B) Impact of Retry impact on SSD's architectures**. As shown in Sect. 4.3.1 (A), the performance degradation caused by uncorrectable pages is linked to the memory access time and to the ECC correction time, which turned out to be negligible in normal read conditions. This implies that the ECC engine is the module to address if we want to reduce the read bandwidth degradation at End of Life (EoL). Because of this, we repeated the analysis by modifying the number of

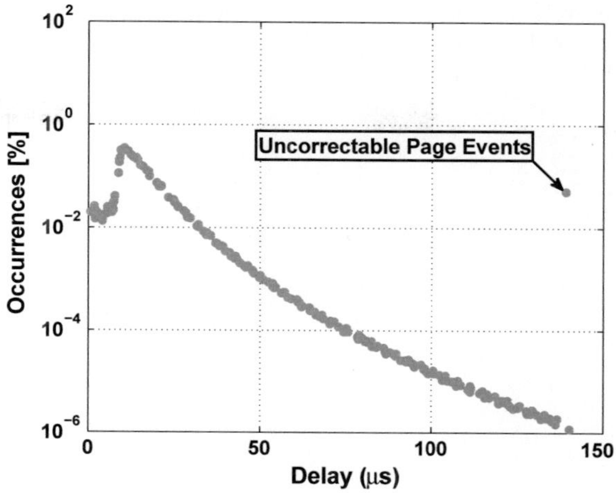

Fig. 4.17 Distribution of the ECC correction times at point B of Fig. 4.14

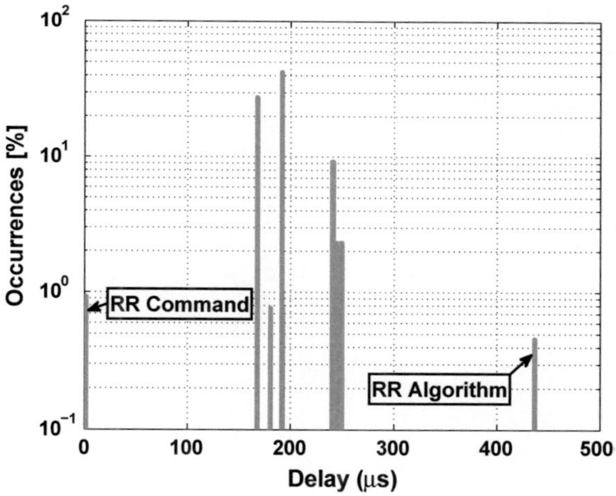

Fig. 4.18 Distribution of the NAND memory latencies at point B of Fig. 4.14

channels and, therefore, the number of ECC engines. Figure 4.19 shows the read bandwidth as a function of RBER, up to a 10% degradation with respect to BoL, for the four channels/NAND flash targets configurations described in Table 4.2.

As expected, by increasing the number of channels (in particular architectures C3 and C4) it is possible to achieve higher bandwidths. The RBER corresponding to a 10% degradation, however, is in the same order of magnitude for the four cases. Therefore, merely increasing the architectural resources, without considering the host interface requirements, does not seem to solve the problem of the RR impact on SSD performances.

When considering the SSD system as a whole, it is clear that the overall performance is driven by the slower between the ECC/memory sub-system and the host interface. As such, if the host interface bandwidth is over designed with respect to that of the ECC/memory sub-system, the performance drop caused by both memory

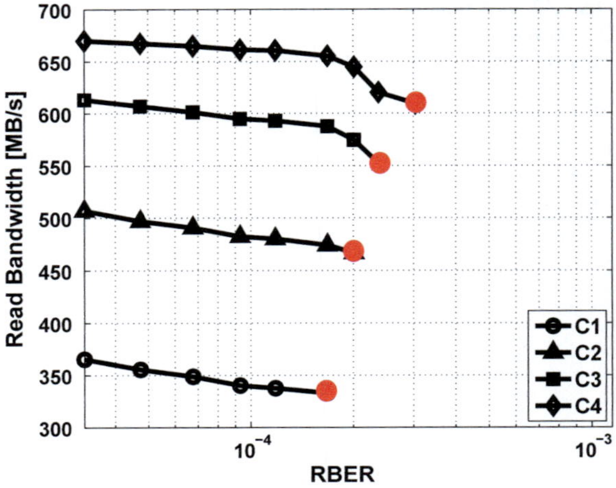

Fig. 4.19 SSD read bandwidth versus RBER for the four architectures described in Table 4.2. *Curves* are shown up to a 10% performance degradation with respect to BoL

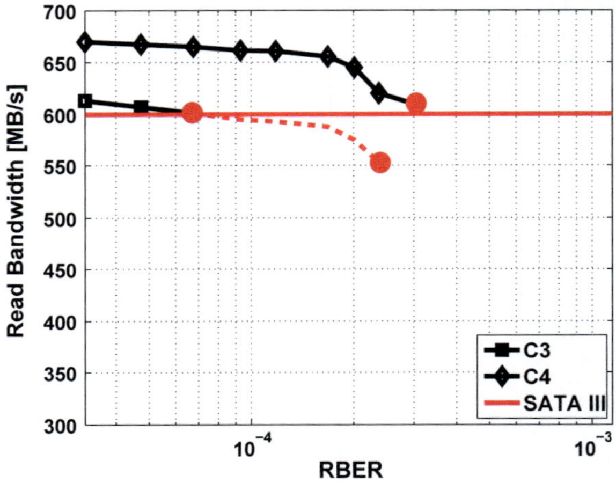

Fig. 4.20 SSD read bandwidth versus RBER for C3 and C4 architectures against the SATAIII theoretical bandwidth. Configuration C4 allows sustaining a larger wear-out prior exposing the performance degradation induced by RR to the user

wear-out and RR techniques can't be hidden to the user. If, on the contrary, the host interface bandwidth represents the system bottleneck, the ECC/memory system must be designed so that, without any waste of silicon (i.e. increase of Flash channels and/or targets) the performance drop caused by memory wear-out and RR techniques does not bring the ECC/memory bandwidth below the one of the host interface. For example, by considering a SATA III host interface, in Fig. 4.20 it is shown that the configuration C3 barely masks the RR-induced performance drop, whereas SSD configuration C4 allows sustaining a higher wear-out level of the NAND, before exposing a "visible" read bandwidth drop to the SSD user.

4.3.2 LDPC Soft Decision

The RBER of Flash memories is exponentially growing because of the aggressive technology scaling needed to increase the memory capacity. Such an increase translates into the inability of reading back the correct data after a number of P/E operations or after long retention times. Figure 4.21 shows the measured average RBER as a function of endurance for three MLC and one • Three-Level Cells (TLC) NAND flash memories manufactured in 2x nm, 1x nm, and mid-1x nm technology nodes as described in Table. 4.3. As the number of P/E cycles increases, the error rate quickly grows up. In addition, either by scaling from a 2x nm to a mid-1X nm node or by switching from MLC to TLC, the RBER increases significantly.

To extend the Flash reliability beyond the rated endurance and, consequently, the lifetime of the drive, the simple combination of RR algorithms and BCH is running out of steam. As a matter of fact, Read Retry can't be used in a massive way: in Sect. 4.3.1 we showed that a percentage of uncorrectable pages in the range of 0.1%

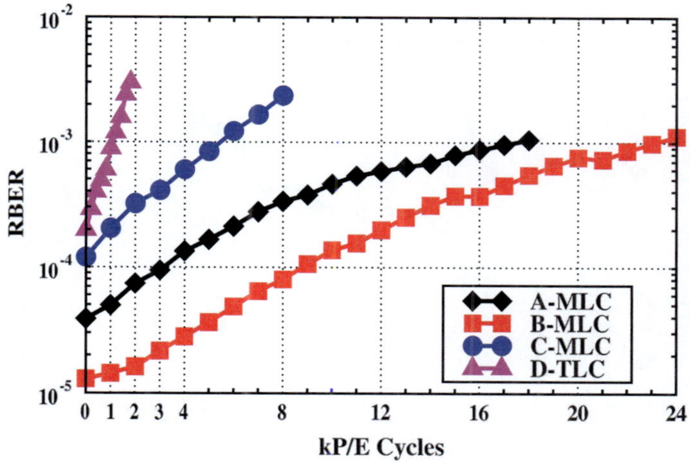

Fig. 4.21 Measured average RBER for 2x nm, 1x nm and mid-1x nm MLC and TLC NAND flash memories as a function of the number P/E cycles

Table 4.3 Main characteristics of the tested NAND flash memories

Sample	Memory type	Rated endurance	Measured average read time (μs)	Measured average program time (μs)	Technology node
A-MLC	Consumer	9 k P/E	68	1400	2X
B-MLC	Enterprise	12 k P/E	40	2000	1X
C-MLC	Enterprise[a]	4 k P/E	70	2500	Mid-1X
D-TLC	Enterprise[a]	0.9 k P/E	86	2300	Mid-1X

[a]Early samples

is already enough to degrade SSD's performances. Figure 4.22 plots the percentage of uncorrectable pages of the curves of Fig. 4.21: the reader can notice how quickly this percentage grows up. Table 4.4 shows the maximum number of Program/Erase cycles that the tested memories can handle before exhibiting uncorrectable pages. Therefore, it is clear that, for extending the endurance capacity of flash, a more sophisticated and much performing ECC is the only practical way to go.

Due to their superior error correction capabilities and decoding performance, *Low Density Parity Check* (LDPC) codes represent a common choice in modern SSDs [22, 23]. Conventional LDPC decoders, if properly designed, can sustain a NAND Flash RBER up to 10^{-2} [23–26]. The LDPC correction is usually performed in two sequential steps: *Hard Decoding* (HD) and *Soft Decoding* (SD), which is called only if HD fails. HD attempts to correct errors starting from a single read operation of the addressed memory page. On the contrary, SD asks for multiple reads.

As summarized in Fig. 4.23, each soft-level requires two page read operations with two different read reference voltages and two data transfers to the ECC engine. The algorithm continues this process until either the page is read correctly or the maxi-

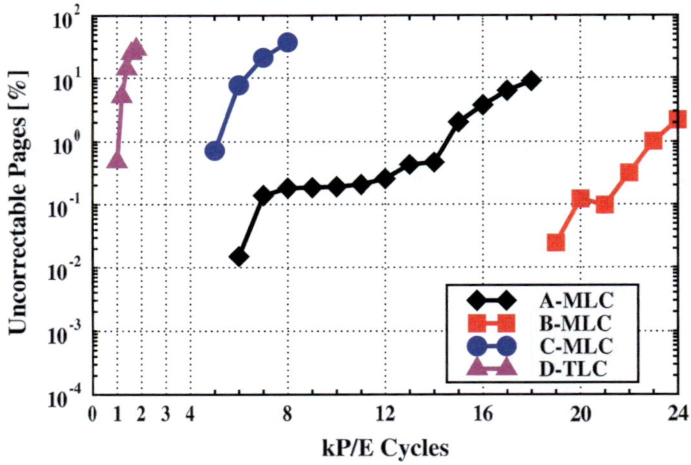

Fig. 4.22 Measured percentage of uncorrectable pages as a function of the number of P/E cycles up to twice the rated endurance. Correction is based on 100 bit/4320 Bytes BCH code

Table 4.4 Maximum number of program/erase cycles without uncorrectable pages for the flash memories of Table 4.3. Correction is based on 100 bit/4320 Bytes BCH code and read retry is OFF

Sample	Measured endurance
A-MLC	6 k P/E
B-MLC	19 k P/E
C-MLC	5 k P/E
D-TLC	1 k P/E

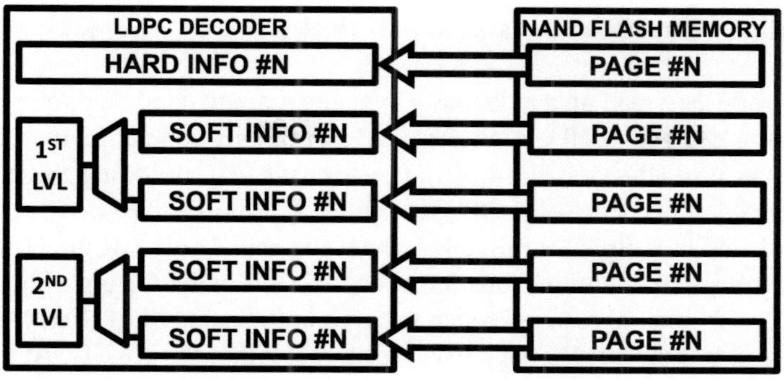

Fig. 4.23 Standard LDPC decoding (HD + 2-level-SD). Each soft-level requires two extra read operations and two data transfer operations

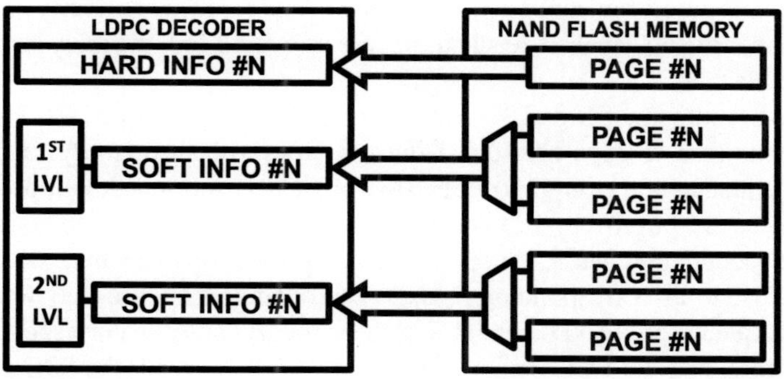

Fig. 4.24 NASD LDPC decoding (HD + 2-level-NASD). Each soft-level requires two extra read operations and only one data transfer, since the 2 read results are combined together inside the NAND chip itself before the data transfer

mum number n of soft-levels is reached. When SD fails, then the page is marked as uncorrectable (unless a RAID correction is available). The overall n-level SD algorithm requires $2n$ page reads and $2n$ data transfers operations. HD has to recover most of the pages without degrading the Code Rate [27]. Hence, HD has a RBER coverage very similar to BCH [26]. Therefore as soon as this strategy fails to correct data, it is requested the intervention of the SD, with a higher correction range. However, in [26] it has been shown that, as soon as the HD approach starts to fail, there is an overhead both in terms of increased SSD power consumption and overall SSD latency, because additional read operations are required (per the SD definition).

An alternative LDPC correction approach that limits the above mentioned drawbacks of SD was described in [28]. The basic idea of this methodology, named *NAND-Assisted Soft Decision* (NASD), is that data for soft decoding are directly produced by the NAND flash memory itself, which internally reads the target page twice for each soft-level. Then, read data are combined together and only one data transfer per soft-level happens, as shown in Fig. 4.24.

Soft Decision Versus NAND-Assisted Soft Decision

Flash memories are read on a page-basis by using a pre-defined read reference voltage, hereafter denoted as HD_0. Cells are read as 1 or 0 by comparing their threshold voltage V_T against HD_0 (see Fig. 4.25a). If, during the ECC decoding phase, the page is uncorrectable, then the Soft LDPC decoding algorithm is called. To accomplish this second step, additional information about the actual value of the NAND flash threshold voltage distributions must be collected. Basically, the algorithm sequentially moves the internal read reference voltages to SD_{10} and SD_{11} (Fig. 4.25b), thus reading the page twice: the 2 page data are stored within the NAND. At this point, these 2 pages are combined in a single page, which is then transferred, byte by byte, to the LDPC decoder, already fed with the page read at HD_0. If the decoding process fails again, a second iteration at SD_{20} and SD_{21} takes place, as shown in Fig. 4.25c. The algorithm continues until either the page is read correctly or the maximum number of soft-levels is reached. In case of overall fail, the page is marked as uncorrectable.

Table 4.5 compares SD and NASD in terms of required operations. NASD can halve the number of page transfers from the NAND flash memory to the ECC. As a consequence, the overall soft decision process is shortened and, hence, SSD performances are improved. To understand the actual NASD efficiency, it must be taken into account that, while the drive is running, not always the page data transfer can stay close to its corresponding NAND read operation. Figure 4.26 sketches the commands queue for NAND dies sharing the same I/O bus, the corresponding data bus allocation, and the ECC engine activity. After a HD_0 read, the SSD controller can send other read or write commands to the same NAND flash die or to other dies. When the ECC engine communicates the read failure to the controller, additional SD_{10} and SD_{11} reads are scheduled. In the SD approach the two Flash pages are transferred separately, as soon as the I/O bus is available; of course, there is a risk

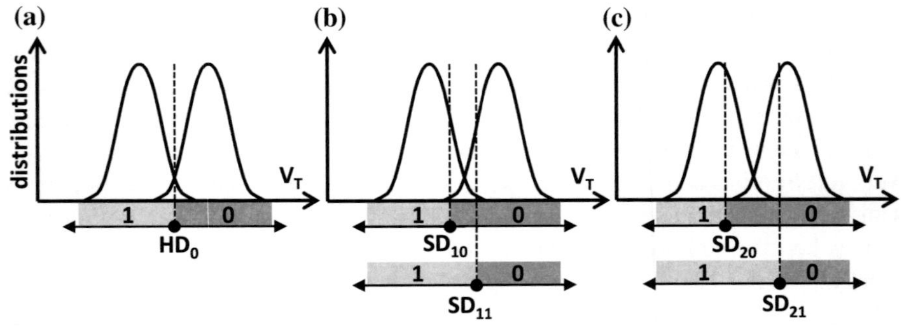

Fig. 4.25 Two levels LDPC sensing scheme. A memory page is read by setting the read voltage at HD_0 and determining, for each bit, whether $V_T < HD_0$ or $V_T > HD_0$ (**a**). If the ECC engine is not able to correct all errors, then the soft decision algorithm kicks in and the page is read by moving the read reference voltages around HD_0, at SD_{10} and SD_{11} (**b**). If the page stays uncorrectable, then other reads at SD_{20} and SD_{21} take place (**c**)

Table 4.5 Read and data transfer operations for SD and NASD approaches

LDPC	One soft-level	Two soft-levels	#n soft-levels
SD	2 page read	4 page read	#2*n* page read
	2 data transfer	4 data transfer	#2*n* page transfer
NASD	2 page read	4 page read	#2*n* page read
	1 data transfer	2 data transfer	#*n* page transfer

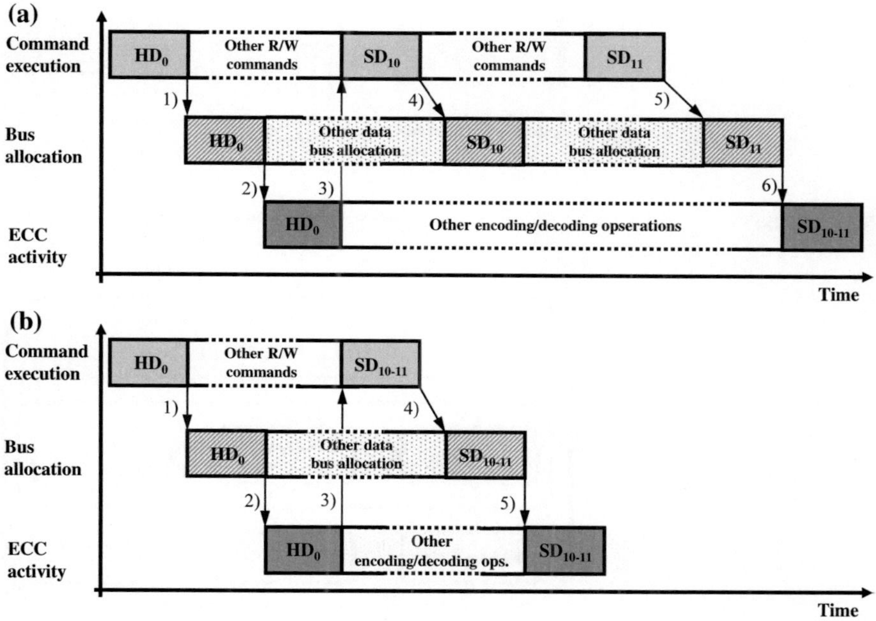

Fig. 4.26 Command queue, data bus allocation, and ECC engine activity for a cluster of NAND flash dies sharing the same data bus. A single soft-level decision operation is sketched. Cases **a** and **b** refer to SD and NASD, respectively

that, between SD_{10} and SD_{11} transfers, the bus is locked by other data transactions to/from other NAND flash dies (see Fig. 4.26a). In the NASD approach, on the contrary, since the outputs of SD_{10} and SD_{11} are combined in a single data transfer, the subsequent soft decision operation can start sooner than in the SD case (see Fig. 4.26b). This advantage becomes even more important when additional soft-levels are required. Furthermore, since the number of data transfers between memory and ECC is reduced, a considerable power reduction of the I/O buses is achieved.

In order to make NASD work, NAND needs to be capable of storing the results of 2 read operations per soft-level. Modern NAND flash devices contain multiple registers because of advanced operations like cache read, read retry, and multi-level storage [29–32]. The NASD implementation can re-use the existing registers inside the NAND: XOR or XNOR gates can then merge the 2 soft reads into a single output

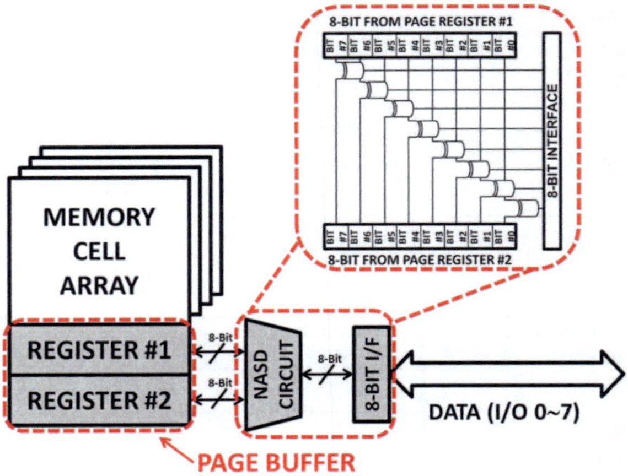

Fig. 4.27 NASD circuitry: 8 XOR gates (or XNOR) have to be added before the 8-bit I/O interface

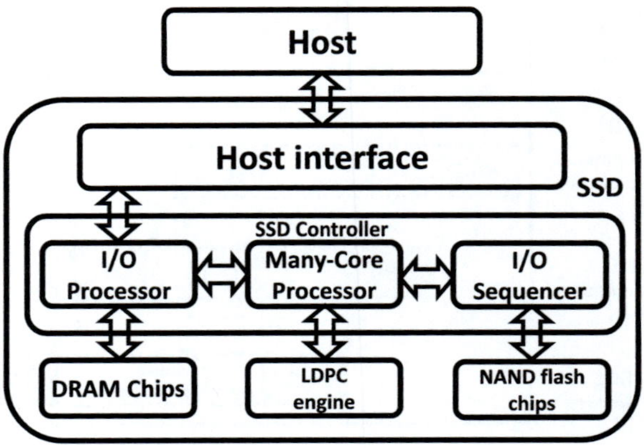

Fig. 4.28 SSD's architecture used to compare NASD versus SD

stream. Since the I/O interface is limited to 8-bits, the logic combination between the pages stored inside the two sets of registers can be performed on-the-fly, byte by byte, during the data transfer phase (see Fig. 4.27). Considering that we are talking about simple logic gates, the NASD implementation inside a NAND flash memory does not impact area and power consumption.

In order to test the effectiveness of NASD with respect to the standard SD, a 512 GB SSD made of 8 channels with 8 NAND flash dies per channel has been simulated. The simulations have been performed considering the 4 different memories described in Table 4.3. Figure 4.28 sketches the main building blocks of the simulated SSD. Besides the standard I/O Processor used for the host-interface address fetch phase, and the many-core Processor busy with the Flash Translation Layer, there is also an I/O Sequencer acting as a read/write scheduler. In order to fully exploit the internal parallelism of an SSD, host random addresses, which could cause die collisions (i.e. requests going to the same die) are parsed and sequentially issued to NAND flash chips.

Table 4.6 Tested host system configurations

	Consumer	Enterprise [33]
Host processor	Intel-Core i5-4570	Intel-Xeon e5-2630
Processor clock	3.2 GHz	2.3 GHz
#N cores	4	24
DRAM size	12 GByte	16 GByte
Workload generator [34]	fio 2.1.10	fio 2.1.10
Avg. I/O submission time	3.5 μs	0.5 μs
Host queue depth [20]	64	256
Host kIOPS (requested)	$\approx$ 200	$\approx$ 600
SSD kIOPS (sustained)	$\approx$ 450	$\approx$ 450

All data have been collected by simulating two different host platforms (Table 4.6). The first one is a consumer system which does not exploit the full SSD architecture (able to sustain 450 kIOPS) since I/O requests settle around 200 kIOPS. As a consequence, all internal error recovery techniques which require additional read operations can be partially hidden. The second one is an enterprise workstation designed to serve hundreds of parallel processes: IOPS demand can go as high as 600,000. In this case the drive cannot meet the target number of IOPS and any additional activity triggered by the error recovery flow hits performances. Thanks to these two different test-cases it has been possible to test the NASD effectiveness when drive resources, such as NAND-flash I/O buses, are partially or completely allocated for user operations.

Results presented in Sect. 4.3.2 (A) refer to an enterprise host and a 100% 4 kB random read workload, which represents the most challenging situation for SSD's performance characterization. In fact, when mixed read/write workloads are considered, since the DRAM chip inside the SSD caches all the write operations, the measured average latency and bandwidth figures of the drive do not reflect the actual SSD behavior. Section 4.3.2 (B) will extend the discussion to realistic workloads for both hosts.

Section 4.3.2 **(A) 100% random read workload—Enterprise host**. Figure 4.29 shows the SSD's read bandwidth gain achieved by NASD versus SD, as a function of the memory endurance. The number of IOPS has been calculated as the average number of read commands completed in a second. For all the considered memories, the NASD technique provides a significant gain. NASD advantages are more pronounced when a large number of uncorrectable pages, triggering a massive soft decoding, is detected.

Figure 4.30 shows the average read latency gains achieved by NASD with respect to SD as a function of memory endurance. Latency has been calculated as the average time elapsed between a read command submission and its completion. All results concerning average latency reflect those obtained for bandwidth (Fig. 4.29).

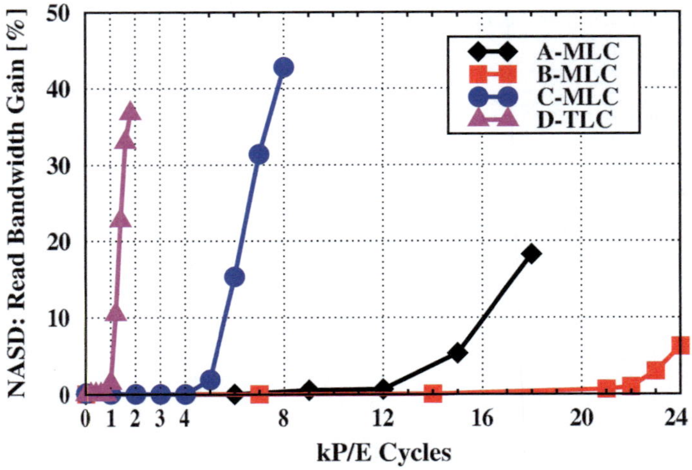

Fig. 4.29 SSD read bandwidth gain achieved by NASD with respect to SD as a function of the memory endurance for the 4 considered memory types and the enterprise host

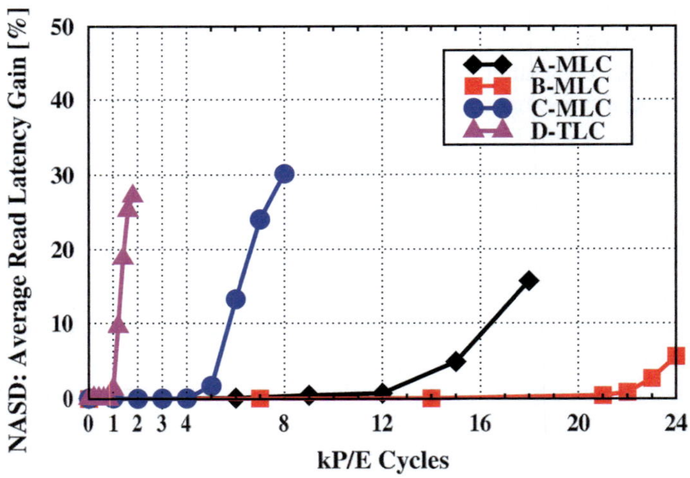

Fig. 4.30 SSD average read latency gain achieved by NASD versus SD, as a function of the memory endurance for the 4 considered memory types and the Enterprise host

Figure 4.31 shows the SSD's cumulative latency distributions calculated at twice the rated endurance for D-TLC, with both SD and NASD approaches. From these data it is possible to extract the SSD's QoS defined as the 99.99 percentile of the cumulative latency distribution [4]. QoS represents the predictability of low latency and consistency of high bandwidth while servicing a defined workload and it can be considered as the key metric to assess the SSD's performance in a worst-case

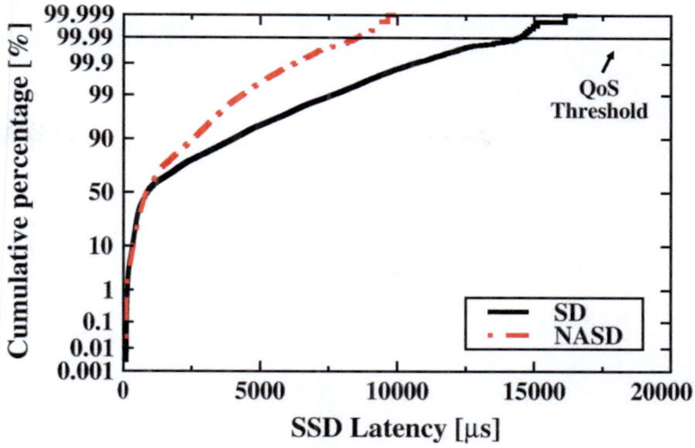

Fig. 4.31 Normal probability paper of the SSD's latency calculated at twice the rated endurance of D-TLC, with both SD and NASD. The QoS threshold is calculated as the 99.99 percentile of the cumulative distribution [4]

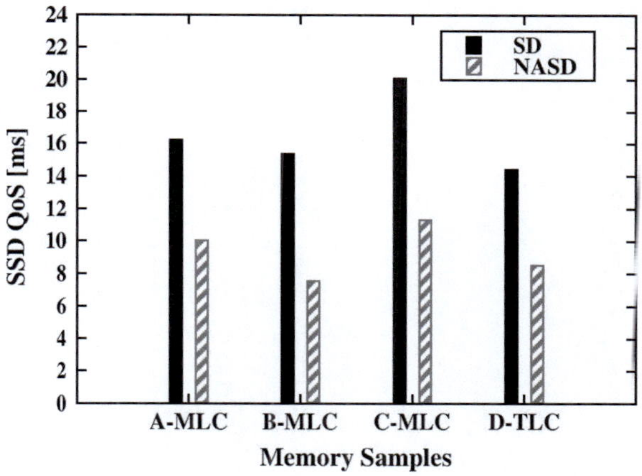

Fig. 4.32 Calculated QoS at twice the rated endurance for the 4 considered memory types and the enterprise host

Table 4.7 Workloads characteristics

Workload	Write ratio (%)	Write amplification factor
MSN	96	1
Financial	81	1.32
Exchange	46	1.94

scenario. Figure 4.32 shows the calculated QoS at twice the rated endurance for all the considered memories, with both SD and NASD approaches.

Section 4.3.2 **(B) Realistic workloads—Enterprise and Consumer hosts**. Since the NASD advantages are tightly coupled to the command pattern, in addition to the 100% random read workload, simulations have also been performed considering three realistic workloads [35], as detailed in Table 4.7; *write ratio* represents the percentage of write commands in the command sequence, whereas *write amplification factor* denotes the number of additional writes produced by the SSD firmware for each single host write [36].

Table 4.8 Bandwidth (in kIOPS for SD and in % of gain for NASD vs. SD) @ twice the rated endurance for both the consumer and the enterprise host

Workload	Consumer host								Enterprise host							
	A-MLC		B-MLC		C-MLC		D-TLC		A-MLC		B-MLC		C-MLC		D-TLC	
	SD	NASD	SD	NASD	SD	NASD	SD	NASD	SD	NASD	SD	NASD	SD	NASD	SD	NASD
MSN	143	4.78	147	4.17	142	5.26	141	5.23	142	4.14	148	4.79	142	5.94	142	5.72
Financial	135	1.45	142	0.54	101	3.68	104	4.36	143	1.88	148	0.45	119	3.67	118	4.19
Exchange	143	2.25	151	0.60	94	4.95	97	5.40	171	2.47	187	0.44	126	5.28	127	6.0
100% read	204	0.03	204	0.03	140	24.41	127	24.85	299	18.24	402	6.24	152	42.77	156	36.80

Table 4.9 Average latency (in μs for SD and in % of gain for NASD vs. SD) @ twice the rated endurance for both the consumer and the enterprise host

Workload	Consumer host								Enterprise host							
	A-MLC		B-MLC		C-MLC		D-TLC		A-MLC		B-MLC		C-MLC		D-TLC	
	SD	NASD	SD	NASD	SD	NASD	SD	NASD	SD	NASD	SD	NASD	SD	NASD	SD	NASD
MSN	373	2.24	343	0.11	393	5.46	396	3.57	1485	0.57	1468	0.16	1553	1.78	1554	1.92
Financial	467	1.50	442	0.53	624	3.60	610	4.29	1724	1.61	1651	0.19	2088	3.99	2093	4.41
Exchange	442	2.21	417	0.60	673	4.84	646	5.21	1465	2.28	1343	0.47	1979	5.28	1973	5.89
100% read	312	0.01	311	0.01	454	19.71	502	19.90	834	15.6	623	5.58	1654	30.14	1620	27.15

Table 4.10 Quality of Service (in ms for SD and in % of gain for NASD vs. SD) @ twice the rated endurance for both the consumer and the enterprise host

Workload	Consumer host								Enterprise host							
	A-MLC		B-MLC		C-MLC		D-TLC		A-MLC		B-MLC		C-MLC		D-TLC	
	SD	NASD	SD	NASD	SD	NASD	SD	NASD	SD	NASD	SD	NASD	SD	NASD	SD	NASD
MSN	47.07	33.95	32.06	36.32	25.17	22.24	34.58	20.46	53.88	28.32	35.48	34.11	31.62	22.73	45.24	22.34
Financial	14.26	17.85	11.43	22.56	12.36	14.59	13.59	13.00	92.60	37.58	80.65	32.61	71.83	5.25	80.75	26.16
Exchange	7.50	21.35	5.93	14.04	9.37	15.11	8.76	14.00	44.20	22.45	37.83	29.53	39.89	16.46	44.21	23.69
100% read	1.50	23.08	0.77	20.85	2.67	21.47	1.59	29.42	16.20	38.10	15.34	50.80	20.08	43.56	14.40	40.84

Tables 4.8, 4.9 and 4.10 show the bandwidth, the average latency, and the QoS, at twice the rated endurance, for the 4 tested NAND Flash memories and for the two host architectures. Simulation results indicate that NASD outperforms SD when other commands are scheduled between the two data transfers required by the SD technique. When realistic workloads are considered, NASD advantages are evident, especially with the MSN workload, which is characterized by a high number of program operations whose duration is significantly longer than read. The QoS improvements for the MSN workload are in a 20% $\div$ 40% range. To sum up, NASD advantages can be clearly identified when QoS is important; in fact, QoS takes into account the worst-case latency conditions rather than the average behavior (i.e. bandwidth and average latency).

References

1. An Overview of SSD Write Caching. http://community.spiceworks.com/attachments/post/0013/5918/ssd_write_caching_tech_brief_lo.pdf.
2. Intel X18-M X25-M SATA Solid State Drive. Enterprise Server/Storage Applications. http://cache-www.intel.com/cd/00/00/42/52/425265_425265.pdf.
3. Samsung NAND Flash memory K9XXG08UXM series. http://www.arm9board.net/download/fl6410/datasheet/k9g8g08.pdf.
4. Intel solid-state drive dc s3700 series - quality of service., 2013. http://www.intel.com/content/www/us/en/solid-state-drives/ssd-dc-s3700-quality-service-tech-brief.html.
5. The cost of latency, 2015. http://www.telx.com/blog/the-cost-of-latency/.
6. Platform as a service (paas), 2015. http://searchcloudcomputing.techtarget.com/definition/Platform-as-a-Service-PaaS.
7. R. Micheloni, A. Marelli, and K. Eshghi. *Inside Solid State Drives (SSDs)*. Springer, 2012.
8. N. Mielke, T. Marquart, Ning Wu, J. Kessenich, H. Belgal, Eric Schares, F. Trivedi, E. Goodness, and L.R. Nevill. Bit error rate in NAND Flash memories. In *IEEE International Reliability Physics Symposium (IRPS)*, pages 9–19, 2008.
9. Sungjin Lee, Taejin Kim, Kyungho Kim, and Jihong Kim. Lifetime Management of Flash-based SSDs Using Recovery-aware Dynamic Throttling. In *USENIX Conference on File and Storage Technologies, (FAST'12)*, 2012.
10. Ren-Shuo Liu, Chia-Lin Yang, and Wei Wu. Optimizing NAND Flash-Based SSDs via Retention Relaxation. In *USENIX Conference on File and Storage Technologies, (FAST'12)*, 2012.
11. Laura M. Grupp, John D. Davis, and Steven Swanson. The Bleak Future of NAND Flash Memory. In *USENIX Conference on File and Storage Technologies, (FAST'12)*, 2012.
12. Yu Cai, Erich F. Haratsch, Onur Mutlu, and Ken Mai. Threshold voltage distribution in MLC NAND flash memory: Characterization, analysis, and modeling. In *Design, Automation Test in Europe Conference (DATE), 2013*, pages 1285–1290.
13. Xueqiang Wang, Guiqiang Dong, Liyang Pan, and Runde Zhou. *Error Correction Codes and Signal Processing in Flash Memory, Flash Memories*. Igor Stievano (Ed.), 2011.
14. Intel-Corporation Robert Frickey. Data Integrity on 20nm SSDs. In *Flash Memory Summit*, 2012.
15. A.G. Cometti, L.B. Huang, and A. Melik-Martirosian. Apparatus and method for determining a read level of a flash memory after an inactive period of time, February 4 2014. US Patent 8,644,099.
16. Silicon Motion Jeff Yang. High-Efficiency SSD for Reliable Data Storage Systems. In *Flash Memory Summit*, 2012.

17. Youngjoo Lee, Hoyoung Yoo, Injae Yoo, and In-Cheol Park. 6.4gb/s multi-threaded bch encoder and decoder for multi-channel ssd controllers. In *IEEE International Solid-State Circuits Conference Digest of Technical Papers (ISSCC)*, pages 426–428, Feb 2012.
18. Intel-Corporation Ravi Motwani. Exploitation of RBER Diversity across Dies to Improve ECC Performance in NAND Flash Drive. In *Flash Memory Summit*, 2012.
19. Serial ATA International Organization. *SATA revision 3.0 specifications*. www.sata-io.org.
20. Nvm express 1.1 specification, 2013. http://nvmexpress.org/wp-content/uploads/2013/05/NVM_Express_1_1.pdf.
21. Pci express base 3.0 specification, 2013. http://www.pcisig.com/specifications/pciexpress/base3/.
22. E. Yeo. An LDPC-enabled flash controller in 40nm CMOS. In *Proc. of Flash Memory Summit*, Aug. 2012.
23. X. Hu. LDPC codes for flash channel. In *Proc. of Flash Memory Summit*, Aug. 2012.
24. Erich F. Haratsch. LDPC Code Concepts and Performance on High-Density Flash Memory. In *Proc. of Flash Memory Summit*, Aug. 2014.
25. Tong Zhang. Using LDPC Codes in SSD — Challenges and Solutions. In *Proc. of Flash Memory Summit*, Aug. 2012.
26. Kai Zhao, Wenzhe Zhao, Hongbin Sun, Xiaodong Zhang, Nanning Zheng, and Tong Zhang. Ldpc-in-ssd: Making advanced error correction codes work effectively in solid state drives. In *Presented as part of the 11th USENIX Conference on File and Storage Technologies (FAST 13)*, pages 243–256, 2013.
27. R. Micheloni, A. Marelli and R. Ravasio. Basic coding theory. In *Error Correction Codes for Non-Volatile Memories*, pages 1–33. Springer-Verlag, 2008.
28. L. Zuolo, C. Zambelli, P. Olivo, R. Micheloni, and A. Marelli. LDPC Soft Decoding with Reduced Power and Latency in 1X-2X NAND Flash-Based Solid State Drives. In *IEEE International Memory Workshop (IMW)*, pages 1–4, May 2015.
29. D.H. Nguyen and F.F. Roohparvar. Increased nand flash memory read throughput, March 8 2011. US Patent 7,903,463.
30. S.H. Lee, S. Bae, J.N. Baek, H.S. Kim, and S.B. Kim. Method of reading data from a non-volatile memory and devices and systems to implement same, March 28 2013. US Patent App. 13/429,326.
31. N. Shibata, K. Kanda, T. Hisada, K. Isobe, M. Sato, Y. Shimizu, T. Shimizu, T. Sugimoto, T. Kobayashi, K. Inuzuka, N. Kanagawa, Y. Kajitani, T. Ogawa, J. Nakai, K. Iwasa, M. Kojima, T. Suzuki, Y. Suzuki, S. Sakai, T. Fujimura, Y. Utsunomiya, T. Hashimoto, M. Miakashi, N. Kobayashi, M. Inagaki, Y. Matsumoto, S. Inoue, Y. Suzuki, D. He, Y. Honda, J. Musha, M. Nakagawa, M. Honma, N. Abiko, M. Koyanagi, M. Yoshihara, K. Ino, M. Noguchi, T. Kamei, Y. Kato, S. Zaitsu, H. Nasu, T. Ariki, H. Chibvongodze, M. Watanabe, H. Ding, N. Ookuma, R. Yamashita, G. Liang, G. Hemink, F. Moogat, C. Trinh, M. Higashitani, T. Pham, and K. Kanazawa. A 19nm 112.8mm^2 64Gb multi-level flash memory with 400Mb/s/pin 1.8V Toggle Mode interface. In *IEEE International Solid-State Circuits Conference (ISSCC)*, pages 422–424, Feb. 2012.
32. Daeyeal Lee, Ik Joon Chang, Sang-Yong Yoon, Joonsuc Jang, Dong-Su Jang, Wook-Ghee Hahn, Jong-Yeol Park, Doo-Gon Kim, Chiweon Yoon, Bong-Soon Lim, Byung-Jun Min, Sung-Won Yun, Ji-Sang Lee, Il-Han Park, Kyung-Ryun Kim, Jeong-Yun Yun, Youse Kim, Yong-Sung Cho, Kyung-Min Kang, Sang-Hyun Joo, Jin-Young Chun, Jung-No Im, Seunghyuk Kwon, Seokjun Ham, Ansoo Park, Jae-Duk Yu, Nam-Hee Lee, Tae-Sung Lee, Moosung Kim, Hoosung Kim, Ki-Whan Song, Byung-Gil Jeon, Kihwan Choi, Jin-Man Han, Kye Hyun Kyung, Young-Ho Lim, and Young-Hyun Jun. A 64Gb 533Mb/s DDR interface MLC NAND Flash in sub-20nm technology. In *IEEE International Solid-State Circuits Conference (ISSCC)*, pages 430–432, Feb. 2012.
33. Hp z640 workstation, 2015. http://www8.hp.com/h20195/v2/GetDocument.aspx?docname=c04434085.
34. Flexible I/O tester, 2015. http://freecode.com/projects/fio.

35. J. Kim, E. Lee, J. Choi, D. Lee, and S. Noh. Chip-level raid with flexible stripe size and parity placement for enhanced ssd reliability. *IEEE Transactions on Computers*, 2014. to appear on.
36. Xiao-Yu Hu, Evangelos Eleftheriou, Robert Haas, Ilias Iliadis, and Roman Pletka. Write amplification analysis in flash-based solid state drives. In *Proceedings of SYSTOR*, pages 10:1–10:9, 2009.

Chapter 5
Resistive RAM Technology for SSDs

Cristian Zambelli and Piero Olivo

Abstract Resistive RAM (RRAM) technology gathered a significant interest in the last decade for system-on-chip applications in silicon-based CMOS technologies like microcontrollers for wireless sensor nodes in the Internet of Things environment, and automotive electronics. Enterprise storage platforms like high performance *Solid State Drives* (SSD) are considering to adopt this technology as a possible storage medium due to its forecasted high reliability, fast access time, and a number of benefits like the native multi-level capability and bit-alterability. In this chapter, we address the basic principles of the RRAM technology starting from the typical cell structure and the physical mechanisms involved in the storage of the information. Thorough analyses of the operations, as well as of the yield and reliability are presented. Then, a review of the most common integration concepts from the 1T-1R to the forecasted 3D cross-point arrays are presented. A brief investigation of the RRAM architectures will help the reader in understanding the density limitations of this technology compared to decananometer scale multi-bit per cell planar and 3D NAND Flash architectures.

5.1 Introduction

Resistive RAM (RRAM) technology gathered a significant interest in the last decade for system-on-chip applications in silicon-based CMOS technologies like micro-controllers for wireless sensor nodes in the Internet of Things environment, and automotive electronics [1–3].

Being hyped from the media as one of the most promising non-volatile memory technologies to compete and replace traditional NAND Flash, after an era of research and development the RRAMs are now facing the slope of enlightenment phase. Enterprise storage platforms like high performance *Solid State Drives* (SSD)

C. Zambelli (✉) · P. Olivo
Dip. Ing., Università degli Studi di Ferrara, Ferrara, Italy
e-mail: cristian.zambelli@unife.it

P. Olivo
e-mail: piero.olivo@unife.it

R. Micheloni (ed.), *Solid-State-Drives (SSDs) Modeling*,
Springer Series in Advanced Microelectronics 58,
DOI 10.1007/978-3-319-51735-3_5

are considering to adopt this technology as a possible storage medium due to its forecasted high reliability, fast access time, and a number of benefits like the native multi-level capability and bit-alterability.

However, the broad expectations on RRAM call for outstanding solutions able to counter the drawbacks of such a memory. To name a few: the extremely high variability in the cells performance integrated in large array structures, the poor control of the read margins through the programming algorithms, and the reduced parallelization of the read/write operations.

In this chapter, we will address the basic principles of the RRAM technology starting from the typical cell structure and the physical mechanisms involved in the storage of the information. Thorough analyses of the operations, as well as of the yield and reliability are presented. Several algorithmic techniques will be demonstrated to prove their effectiveness in addressing those metrics. A reliability threat assessment in terms of endurance and data retention will be provided with a focus on the physical roots behind the memory lifetime degradation. Then, a review of the most common integration concepts from the 1T-1R to the forecasted 3D cross-point arrays are presented. A brief investigation of the RRAM architectures will help the reader in understanding the density limitations of this technology compared to decananometer scale multi-bit per cell planar and 3D NAND Flash architectures. Finally, a section of this chapter will study the typical disturbs evaluated in RRAM array architectures to observe the operation margins in the typical memory working window and some possible countermeasures to keep them on an acceptable level.

5.2 Basic Principles and Operations of RRAM Cells

The basic principle of the RRAM technology (sometimes quoted as memristive technology [4]) is the so-called resistive switching, namely the peculiar property of a material to change its resistivity as a function of an applied voltage or current bias. The switch is theoretically from an insulator state (i.e., high resistivity) to a metallic-like conduction state (i.e., low resistivity). The change of the conduction regime in the cell is reversible and the two different resistivity states define two well-separated logical states.

As stated in [5], there are a number of resistive switching devices that differ only from the physics mechanisms: nanochemical memories, phase change memories, valence change memories, thermochemical memories, magnetoresistive memories, etc. Among them, the valence change memories are the most attractive concept for RRAM cells integration because of the materials compatibility with the state-of-the-art *Back-End-Of-Line* (BEOL) process scheme. Various suitable materials for RRAM including chalcogenides, perovskite-type oxides, and binary transition metal oxides (e.g. Ta_2O_5, TiO_2, and HfO_2) are promising candidate materials showing excellent switching capabilities [6, 7].

In these memories, the RRAM behavior is based on the possibility of electrically modifying the conductance of a *Metal–Insulator–Metal* (MIM) stack. The MIM is

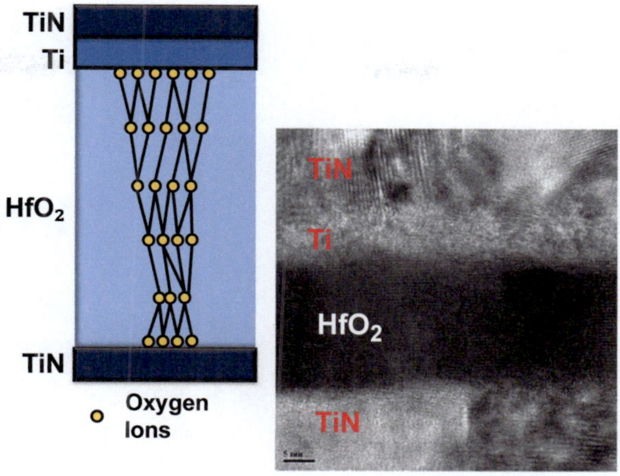

Fig. 5.1 RRAM cell structure (*left*) and STEM image (*right*) considering TiN-based electrodes, Ti oxygen exchange layer, and HfO_2 as switching oxide

composed by two electrodes (i.e., bottom and top electrode), a switching oxide layer, and an oxygen exchange layer, as shown in Fig. 5.1.

A Set operation moves the cell in a *Low Resistive State* (LRS), whereas Reset brings the cell back to a *High Resistive State* (HRS). A read operation, consisting in a small bias voltage applied to the cell to get its resistance value according to the flowing current, is used to discriminate between the two states. The physical basis of this process is ascribed to the formation and rupture of a filamentary conduction path formed in the insulation layer by oxygen ions and vacancies (V_O^+). To initiate such a switching behavior, some technologies require a preliminary forming operation [8–10]. Figure 5.2 depicts a state transitions schematics. The Forming, Set, and Reset parameters like the voltage and the timings required for the operation vary depending on the materials and on the cell structure.

5.2.1 Forming Operation

The Forming process in RRAM cells is performed just once, but this initial state plays a fundamental role in determining the subsequent cells performance [12]. The effectiveness of the forming process depends on its ability in creating a stable conductive filament, thus easing successive Set/Reset operations. The Forming is similar to the soft breakdown process in dielectrics. It is possible to modulate the size of the conductive filament depending on the magnitude of the current flowing in the dielectric, usually controlled by a select device in series to the MIM stack (e.g., a MOS transistor or a diode). Standard Forming in array architectures is performed by applying either a voltage ramp or a voltage/current pulse to each memory cell individually [9]. This technique is indicated as *Incremental Forming* (IF). The former method has a major drawback due to the filament conductance control being not tight enough. This results in larger cell-to-cell variability and larger disturb sensitivity [9, 13]. As an alternative, Forming process can be performed through a sequence of

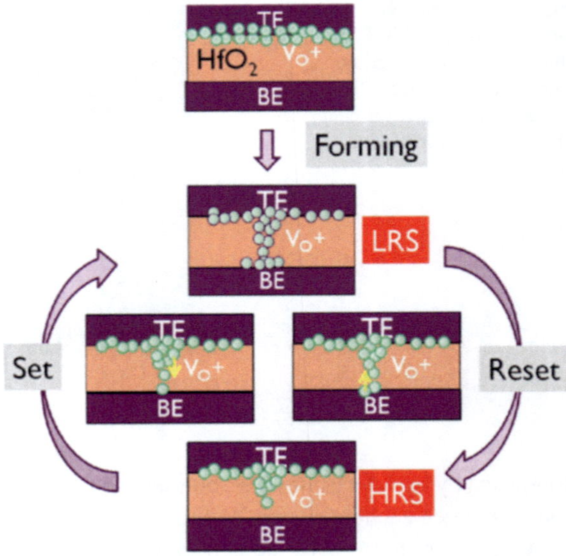

Fig. 5.2 State transitions schematics from the forming operation to cycled set and reset operations [11]

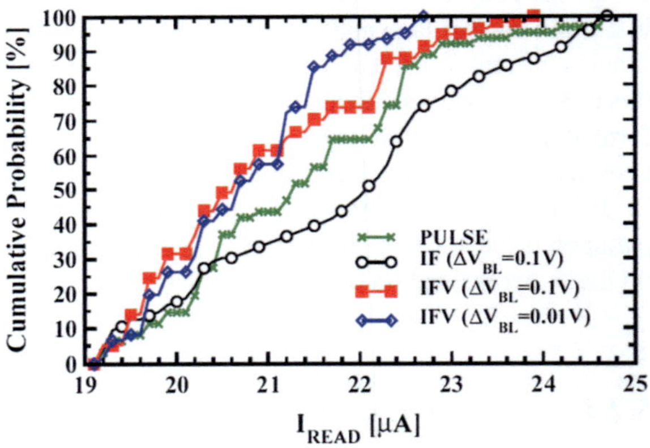

Fig. 5.3 Cumulative read current distributions after Forming (i.e., variability measure) of correctly formed cells (i.e., yield) using the following schemes: Pulse (54%), IF (77%), IFV with ΔVBL = 0.1 V (87%), and IFV with ΔVBL = 0.01 V (99%) [15]

pulses featuring the same voltage. While the Forming time can be minimized using a single pulse with high compliance and voltage parameters, the Forming yield is limited since the applied energy is not sufficient to complete the Forming process in all cells [14]. Several pulse-based forming alternatives have been proposed to increase the applied energy and therefore the yield by using long pulses or sequences of short pulses at constant voltage, and even Form and Verify schemes (IFV) [9, 15]. Figure 5.3 shows a comparison of different Forming schemes in terms of yield (i.e., percentage of cells correctly formed) and in terms of variability control applied to RRAM test arrays.

It is worth to point out that the Forming operation depends upon several technological factors like the area of the MIM cell, the thickness of the switching oxide

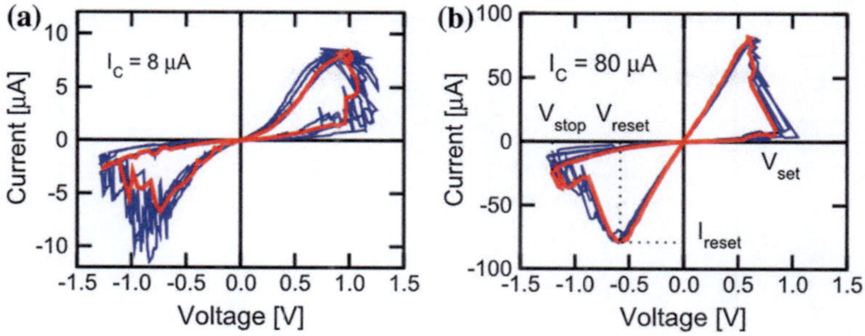

Fig. 5.4 Measured I-V curves showing the variability of the set and reset switching voltages for different compliance currents [18]

layer, and the phase of the insulating material that can be deposited in the MIM stack, either in amorphous or poly-crystalline flavor [16, 17].

5.2.2 *Set and Reset Operation*

The Set and Reset operations in RRAM cells are those deputed to switch between different memory logical states. After the creation of the conductive filament with the Forming operation there is a modulation of its shape due to the transport of oxygen vacancies and defects in the switching oxide. Such behavior depends on the bias voltage polarity applied to the top and bottom electrode of the MIM stack. Usually the Set operation requires a positive bias applied on the top electrode, whereas the Reset operation requires a negative one.

From the physics point of view, the transition from Set to Reset operation corresponds to the switching of the conduction regime of a quantum conduct wire. In the Set state (LRS) the conduction mechanism is metallic-like, whereas in the Reset state (HRS) the conduction is usually associated to a direct tunneling charge transport through a finite potential barrier, whose height and thickness depend on the shape of the conductive filament constriction created during the Reset operation [18].

By applying a voltage sweep to a RRAM cell to measure its I-V characteristics we observe that the Set transition takes place at a positive voltage V_{Set}, whereas the onset of the Reset transition can be seen at a negative voltage V_{Reset} and current I_{Reset}, as indicated in Fig. 5.4 [19]. The negative sweep is generally completed at a negative voltage V_{Stop}, which is necessary to achieve the high resistance of the Reset state. The current compliance I_C, forced by a select device like a MOS transistor, controls the overall MIM resistance in the Set state and the I_{Reset}, thus playing an essential role in limiting the power consumption in the memory cell.

By analyzing the characterizations performed either on a single RRAM cell [19] or on a large number of RRAM cells integrated with the same manufacturing process in an array structure [20] it is noticed a significant variability of the switching voltages

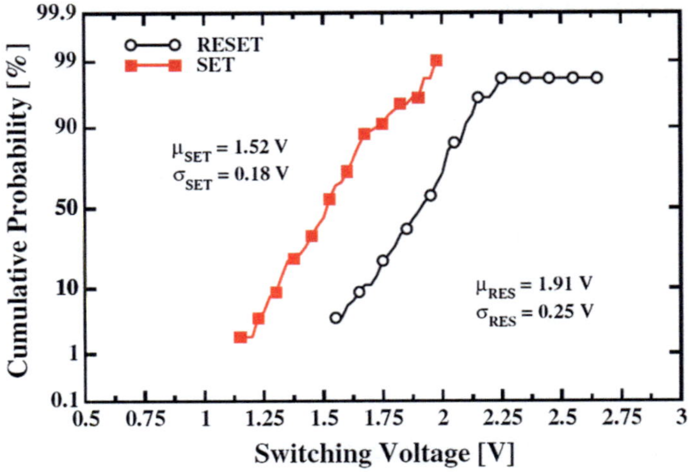

Fig. 5.5 Cumulative set and reset switching voltage distribution retrieved from the characterization of a 4 kbits RRAM array [20]

responsible for Set and Reset operations (see Fig. 5.5). This calls for an optimization of the Set and Reset schemes in order to improve the memory reliability and first of all the switching yield.

Cycling operation performed with non-optimal Set/Reset pulses may lead to early failure, in either LRS or HRS, as shown later in this chapter. The failure mode depends on the "strength" of the Set/Reset pulses [21], relative to each other. Careful tuning of the cycling conditions leads to balanced Set and Reset operations and allows for improved endurance of over 10^{10} cycles for single cell structures, which is comparably higher than what exposed by state-of-the-art NAND Flash technology [22].

5.3 Reliability and Performance of RRAM Cells

Considerable progress has been made in RRAM device integration as well as in understanding the physical/chemical properties of the resistance change behavior. Although several RRAM demonstrators showed excellent performance parameters [23–25], there are some reliability concerns that require a deeper understanding before claiming that RRAM technology will outperform NAND Flash. Among them it is mandatory to study the endurance properties and the data retention capabilities. Moreover, it is important to understand what the limitations of the algorithmic techniques are for Set and Reset operations caused by phenomenon like the Random Telegraph Noise (RTN) in order to improve those metrics. Moreover, endurance and retention are severely impacted by the inter-cell variability (i.e., variations between cells) and the intra-cell variability (i.e., cycle-to-cycle variations of any given cell) that still prevents RRAM mass production and fast commercialization [26]. In this section we will analyze all the single aspects of the performance and the reliability of RRAM cells starting from the basis of the variability reaching up to the most common issues for this technology.

5.3.1 Intra-cell Variability

While in the well-established Flash NAND technology variability mainly shows-up as device-to-device (D2D) fluctuation as a consequence of device scaling, RRAM, due to their fundamentally different operating mechanism, additionally display significant intra-device, cycle-to-cycle (C2C) dispersion [27].

Figure 5.6 shows the I-V characteristics of a single cell measured for different Set and Reset operations. As can be seen, the same cell exhibits a large distribution of the switching voltages, making difficult the usage of single voltage pulse Set and Reset approaches to switch the logical state of a cell. A too high voltage will result in unwanted stresses applied to the cell that switches early, whereas a too low voltage will turn into an incomplete switching transition between the states of slow-switching cells. The origin of the intra-cell variability seems to be ascribed both to the stochastic nature of the switching process during Set and Reset transition and to the Forming operation, therefore being related to the shape of the conductive filament [26].

5.3.2 Inter-cell Variability

The inter-cell variability is captured by monitoring multiple cells in test element groups or array structures to evidence their difference in terms of performance and reliability for a given operation. As for the intra-cell variability, it is believed that this variability source is due to the nature of charge transport in RRAM technology. However, when speaking of inter-cell variability, we must consider two different

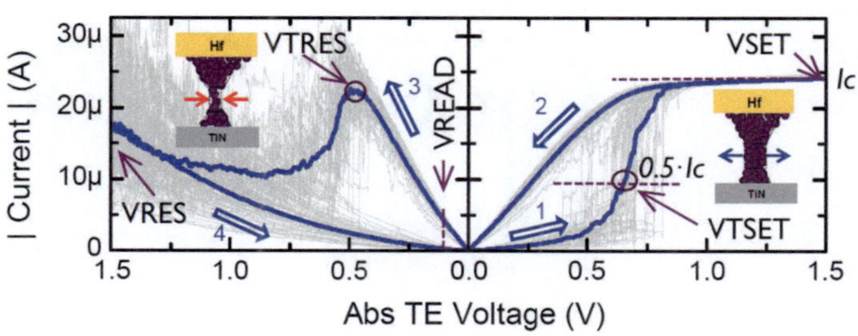

Fig. 5.6 Single RRAM cell DC I-V cycling over 50 cycles at 25 μA current compliance range: gray all cycles, black median values [27]

concepts: the intrinsic variability, due the manufacturing process of the memory cell and to the switching nature, and the extrinsic variability, mainly ascribed to the integration concept used for the cells [28].

Concerning the intrinsic inter-cell variability, the research community looks at the variability introduced in the process steps of the MIM element, which is the core of the RRAM cell. The thickness inhomogeneity of the top and bottom electrodes is not believed to play a central role in the inter-cell variability. However, the combination of the bottom electrode deposition process with the switching oxide deposition process seems to affect the variability especially in the Forming operation (see Fig. 5.7), thus reflecting on the consecutive Set and Reset operations. This has been proven in HfO_2 [26] and in AlO_x RRAM cells [29], where the roughness of the switching oxide films due to the material structure of the electrodes becomes significant.

The inter-cell variability is additionally evaluated for HRS and LRS resistances. In RRAM technology the HRS inter-cell variability is considerably higher than the LRS one, as shown in Fig. 5.8. However, we have to consider that HRS and LRS inter-cell variabilities are strictly related since they are linked to the shape of the conductive filament created with the Forming operation and successively modulated by the cycling Set and Reset operations.

5.3.3 Endurance

The endurance degradation is a critical reliability issue in every non-volatile memory technology, and RRAM makes no exception. Thinking about multi-level cell NAND Flash, the presence of dedicated algorithms to counter the wear-out of the tunnel oxide is mandatory to achieve reliable storage of the information. Same thoughts are applied to RRAM. Indeed, without an accurate study of the wear-out mechanisms in the cells, of the optimization of Set and Reset algorithms, and of the quantification

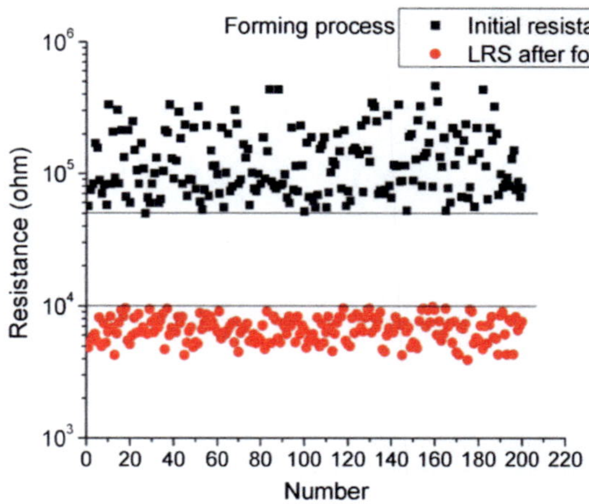

Fig. 5.7 200 AlO_x-based RRAM cells tested after the manufacturing (initial resistance) and after forming operation [29]

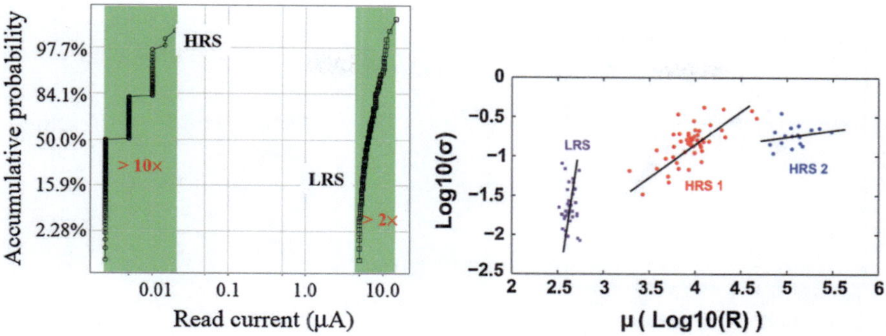

Fig. 5.8 Distribution of read current in LRS and HRS of Cu_2O-based resistive switching devices (*left*) [30] and inter-cell variability analysis in HfO_2 RRAM cells (*right*) [31]

of the cells variability, it is quite impossible to extend the RRAM lifetime up to 10^4-10^5 cycles in array structures.

Figure 5.9 illustrates two kinds of typically measured endurance degradation characteristics considered as "over-Set" and "over-Reset". In the "over-Set"/"over-Reset" endurance degradation, the resistance of the high/low resistance state (R_{HRS}/R_{LRS}) gradual decreases/increases firstly, then sharply decreases/increases, being successively unable to move from LRS/HRS, which causes the loss of the margin between the two logical states. Their corresponding physical mechanisms are also shown in Fig. 5.9 [22, 32, 33]. The "over-Set" endurance degradation is attributed to the extra oxygen vacancies (V_O) generation during Set process, which causes the extra growth of the conductive filament size, together with the reduced R_{HRS} and R_{LRS} [32]. This increased filament size may cause Reset failures [34]. The "over-Reset" originates from extra recombination between V_O and oxygen ion (O^{2-}), which causes the widening of the electron tunneling gap. This may cause Set failure due to the weakening of electric field in gap region, resulting in the reduced V_O generation [34].

In [22] it was proven a methodology to increase the endurance of single RRAM cells by modifying the Set/Reset pulse amplitudes or pulse widths. Compared to varying the Set/Reset pulse amplitude, tuning the Set/Reset pulse width do not significantly change the failure mode of the endurance. Generally, short Reset pulse demonstrates larger influence on the endurance stability, as compared to the Set pulse. By optimally choosing the Set/Reset parameters it was possible to achieve up to 10^{10} endurance cycles in single cell structures. Unfortunately, this tuning approach is not applicable in large scaled RRAM arrays, where the high number of cells increases the impact of the inter-cell variability during cycling. Moreover, a poor tuning of the Set/Reset algorithms will have a dramatic impact on the intra-cell variability and therefore on the overall reliability of the array.

To counter the inter-cell variability, and to ensure a significant margin between HRS and LRS, several Set and Reset verification algorithms have been proposed in literature (see Fig. 5.10 to appreciate the outcome of one of them) to prevent unwanted

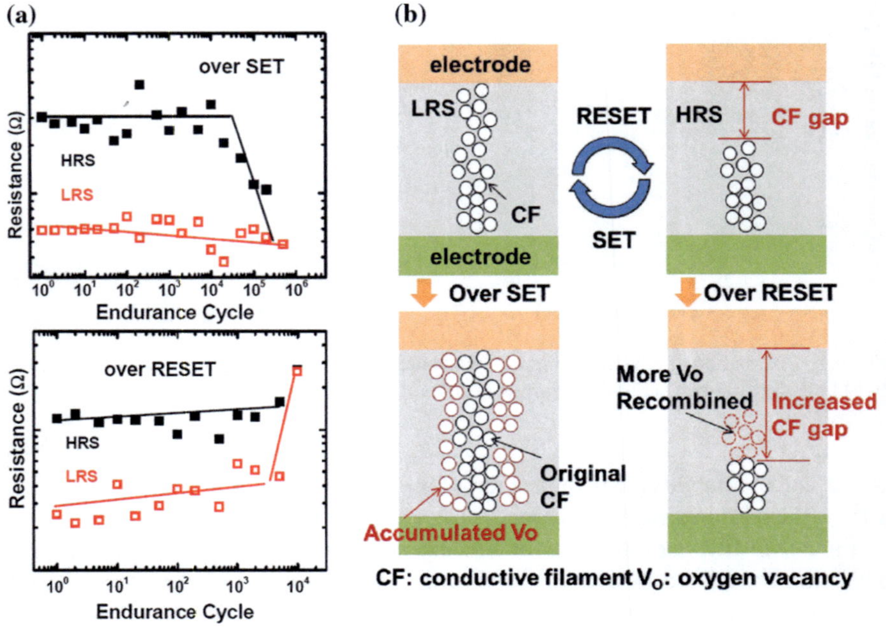

Fig. 5.9 **a** Typical endurance degradations observed in the experiments. **b** Schematic of endurance degradation mechanisms [34]

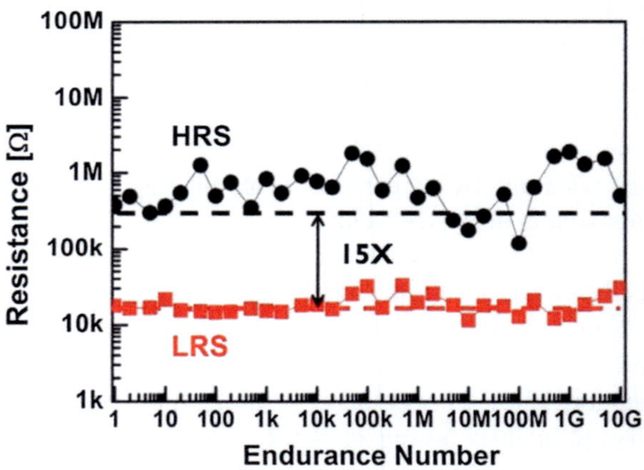

Fig. 5.10 By balancing the set pulse and the reset pulse, 10^{10} pulse endurance was achieved for a single 40-nm Hf/HfO_2 RRAM cell [22]

stress on the memory cells and to maximize the switching yield in array structures [22, 35].

5.3.4 Data Retention

The data retention capabilities of the RRAM technology must be evaluated as for NAND Flash memories by considering different temperature ranges: the *Low Temperature Data Retention* (LTDR) and the *High Temperature Data Retention* (HTDR). Both ranges evidence that Forming, Set, and Reset operations have an influence on the retention properties of the memory.

The LTDR has been evaluated in 50 nm Al_xO_y MIM cells, although the results generally apply to other switching oxides as well [36]. The HRS before Forming and LRS after Forming show the best retention compared with after Set/Reset cycling operation. However, this long retention time without forming is limited to one-time-program (OTP) applications. When the memory endurance is stressed by cycling, the LRS retention becomes better, whereas HRS retention gets worse, because of the larger size of the conductive filament in the switching oxide. Consequently, the retention time of RRAM is degraded as the device is worn out and the error rate of HRS becomes the dominant one.

For the HTDR analysis, several studies have been performed in literature to help understanding the physical nature of the retention loss in RRAM cells [22, 37–39]. Two possible mechanisms responsible for the retention loss (illustrated in Fig. 5.11) have been proposed:

1. mobile oxygen (scavenged by the oxygen exchange layer during the cell stack deposition, post-processing and Forming operation) diffuses back into the switching oxide and recombines with V_O in the filament;
2. V_O out diffusion and dissolution of the filament [38].

The degradation mechanism 1 is area dependent as the source of mobile oxygen participating in degrading the retention is determined by the area of the top electrode. In contrary, the mechanism 2 is related to the changes of the filament shape and hence area independent.

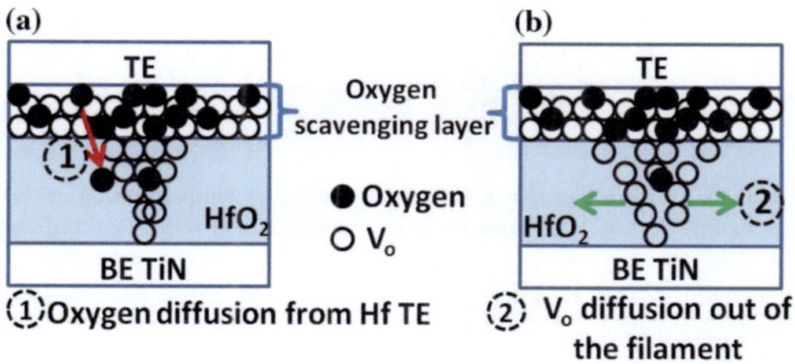

Fig. 5.11 Two possible retention degradation mechanisms in HfO_2 / Hf RRAM cells: 1. **a** oxygen scavenged by Hf cap layer diffuses back into HfO_2 and recombine with V_O in the filament; 2. **b** V_O out diffusion and dissolution of the filament. Mechanism 1 is area dependent in degrading retention while mechanism 2 is area independent [37]

The read-out current of both LRS and HRS baked at 150, 200 and 250 °C decreases with longer bake time, indicating filament continuous "dissolution", and leading to the increase in resistance. Clear temperature dependence of retention degradation is observed: the median LRS and HRS read-out current decreases much less at lower temperature (Fig. 5.12), indicating the temperature activated nature of retention degradation of oxygen vacancy (V_O) filamentary switching. The resistive states degrade much faster in smaller cells. After 120 h, 250 °C baking, the LRS read-out current in 40 nm MIM cells reduces one order of magnitude more than the 320 nm MIM cells. Lowering the operation current down to low compliance values further degrades data retention. Due to the reduced amount of V_O in the filament, its stability is also greatly decreased [37]. By limiting the oxygen diffusion with an additional annealing applied after RRAM cell formation, the retention could be greatly improved.

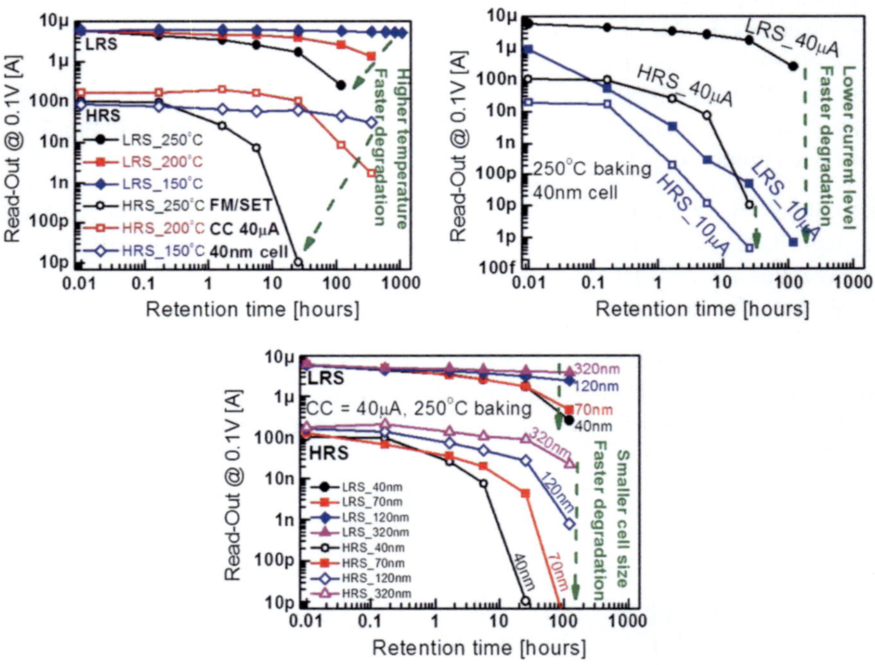

Fig. 5.12 Median HRS/LRS values as a function of the bake time performing retention tests at different temperatures (*top left*), different current compliance (*top right*), and different cell size (*bottom*) [37]

The HTDR considering the inter-cell variability impact has been studied in [22] on array structures by performing 125 °C bake experiments after one Set/Reset cycles and 10 k cycles. The results confirm what observed previously: the temperature impacts the retention capabilities of the memory cells depending on the number of cycles performed, the larger impact is retrieved for HRS, and the inter-cell variability of the Set/Reset operations is maintained after the retention tests.

5.3.5 Random Telegraph Noise and Current Instabilities

The stochastic characteristics of the RRAM can be evidenced not only during the Forming, Set and Reset operations, but also during the Read procedure. This represents a limitation of the operation and verify algorithms since phenomenon like the *Random Telegraph Noise* (RTN) or post-Forming/Set/Reset instabilities lose control of the HRS and LRS distributions.

The RTN is ascribed to the charge-transport through the dielectric barrier in HRS dominated by the *Trap-Assisted Tunneling* (TAT) process, which is supported by activated defects, like oxygen vacancies in a positively charged state: ${V_O}^{2+} + e^- \rightarrow {V_O}^+$. When the vacancy loses an electron (${V_O}^+ \rightarrow {V_O}^{2+}$), its ionization and relaxation energies change, disabling the electron transport via this defect. RTN in HfO_2-based RRAM devices was attributed to such activation and de-activation of the TAT-supporting defects [40]. Furthermore, TAT transport can be also affected by charging and discharging of defects not directly contributing to the electron transfer: capture/emission of an electron by one of such defects may lead to a Coulomb blockade of the nearby TAT-supporting trap, thus changing the capture/emission times of the electrons transit (can either reduce or enhance the TAT conductivity). Figure 5.13 shows a schematic of these processes. In both above mentioned physical mechanisms, the event triggering RTN is the charge capture/emission, which allows proposing a "universal" simplified description of the RTN process [40].

The effect of the RTN (see Fig. 5.14) on the RRAM reliability is to change the efficiency of the verify operation in the incremental pulse algorithms for endurance improvement and inter-cell variability control [15, 35, 41].

Another issue in the incremental pulse algorithms is represented by the post-operation instabilities [42]. Especially at low operating current, the width of the HRS and LRS distributions becomes unacceptably large, resulting in a LRS/HRS distribution overlap at low percentiles with read errors as a consequence. *Incremental Step Pulse* (ISP) algorithms have been proposed to resolve this issue [35], but recently [42] it was demonstrated that, for RRAM devices, post-Set and post-Reset instability of the filament make these algorithms highly ineffective. This is because after the verify step, both the LRS and HRS distributions always evolve towards a wider distribution.

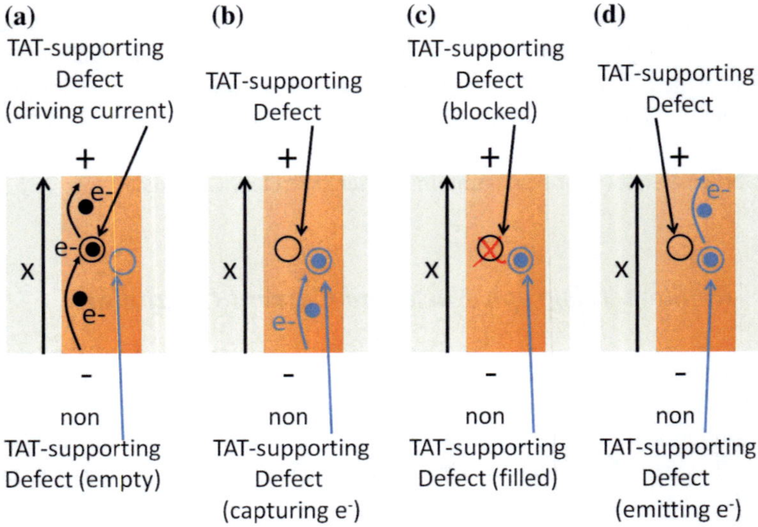

Fig. 5.13 Schematics of a possible RTN-generating mechanism for a RRAM device in HRS. The *dark gray cylinder* represents the dielectric barrier, the thickness of which is 5. **a** The TAT electron transport via an activated defect. **b** The non TAT-supporting defect captures an electron resulting in a **c** Coulomb blockade of the TAT-supporting defect blocking the charge transport through it. **d** The non TAT-supporting defect emits the captured electron restoring the charge transport properties of the TAT-supporting defect properties of the TAT-supporting defect [40]

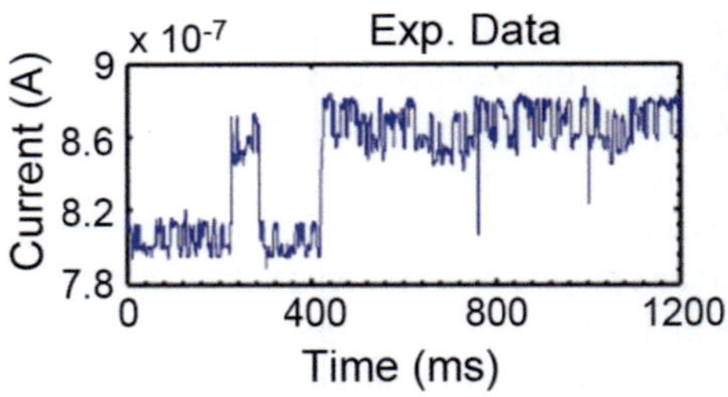

Fig. 5.14 Experimental multi-level RTN in HRS at $V_{READ} = 0.1$ V (HfO_2/Ti 4.2 nm/5 nm) [40]

5.4 RRAM Integration: Architectural Solutions

The characteristics of the RRAM cells in terms of performance and reliability have been largely evaluated on single cell MIM structure that provided a good vehicle to understand the physical limitation of such a technology, yet being unable to assess specific issues like read and write disturbs, typical of integrated array solutions. In this section we review the most common array solutions for RRAM by comparing their pros and cons in terms of offered integration density, performance, and disturbs immunity.

5.4.1 True Cross-Point Arrays

To enable continued scaling, the RRAM true cross-point memory architecture appears as one of the most attractive successors to the current Flash technology, due to its inherent $4F^2$ cell size and simplicity in fabrication [43]. The cross-point memory architecture is composed of a MIM memory element sandwiched by two sets of parallel conductive interconnects crossing perpendicularly, as shown in Fig. 5.15. Wrong Set/Reset and misreading can readily happen in the cross-point architecture due to the substantial sneak path leakage in the half-selected and unselected cells. Indeed, to guarantee a successful write operation to both states, the switching threshold of the selected cell (V_{Select}) must be larger than the maximum values of V_{Set} and V_{Reset} under the worst-case scenario. On the other hand, the voltage drop on the unselected cells should be smaller than the minimum values of V_{Set} and V_{Reset} ($V_{Unselect}$) to eliminate write disturbs. To program a cell, it is used the $V_{dd}/2$ writing scheme [44] in which the selected word line is biased at V_{dd}, the selected bit line is grounded, and all the unselected word lines and bit lines are biased at $V_{dd}/2$ (see Fig. 5.16). During the write operation, an extra voltage drop along the interconnects caused by the leakage current can lead to an insufficient voltage at the selected cell required for a successful write.

There are several ways of reading a cross-point memory cell. The most common one is the bias scheme in which the selected word line is biased at V_{Read}, and all the other word lines and bit lines are grounded. This method maximizes the readout throughput by reading multiple cells at the same time, although the power consumption increases correspondingly. To ensure a nondestructive read, V_{Read} is set to be smaller than $V_{Unselect}$. The current difference (ΔI) when reading HRS and LRS is then sensed by connecting all the bit lines to sense amplifiers. During the read operation, without a select device in series, parasitic conducting paths in unselected cells can degrade the output signal and make it hard to discriminate the two states of a memory cell; therefore, read disturb is a concern for true cross-point RRAM architectures [43, 44].

Concerning the scalability of this integration approach we must note that the maximum array size is only dependent on LRS under the worst-case scenario. Figure 5.17 shows the minimum requirement of LRS as the array size increases. It is clear that

Fig. 5.15 Layout of a true crosspoint RRAM array [44]

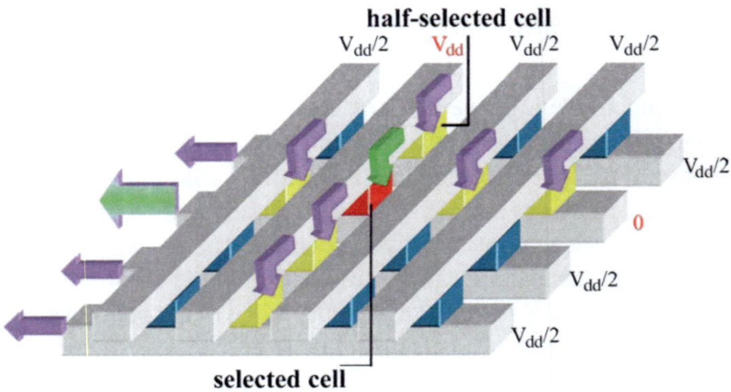

Fig. 5.16 Schematic view of the cross-point memory structure. (*Arrows*) Flow of current paths through the selected and half-selected cells during the write operation. (*Black*) Selected cell. (*Light gray*) Half-selected cells. (*Dark gray*) Those with equal potential at the corresponding word lines and bit lines [43]

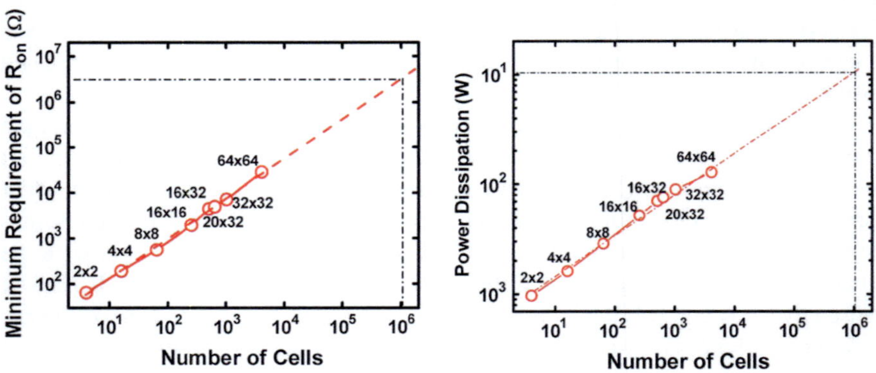

Fig. 5.17 Minimum requirements (*left*) for LRS (R_{on}) and power dissipation (*right*) as a function of the integrated number of RRAM cells [43]

the minimum requirement of LRS increases linearly with the number of memory elements. Increasing LRS from 1 kΩ to 3 MΩ scales up the number of cells in the array to 10^6. In this case, it eliminates the need for a memory cell selection device for cutting off leakage paths and increases array size. High resistance values also reduce the power dissipation in the array, which is one of the most critical issues for memory application. Figure 5.17 shows the increase in power dissipation as the size of the memory array grows. Power consumption of as high as 0.1 W is expected as the number of cells in the array increases up to 10^6, with LRS = 5 kΩ. To reduce the power consumption, the resistance values of the memory cells must be scaled up to make the large memory array feasible. However, it must be noted that high LRS values can degrade the write and read speeds of the memory [43].

5.4.2 1T-1R, 1D-1R, and 1S-1R Arrays

The ideal solution to reduce the disturbs and the scalability issues of the true cross-point architecture is to use a select device like a transistor in series to the RRAM cell. This approach is known as the 1T-1R array (see Fig. 5.18). The transistor element is an ideal selecting element since it provides isolation for the unselected cells in the array and limits the current (through a compliance setting) of the selected cells in Forming, Set, and Reset operations. Since the 1T-1R element is basically a three-terminal device, large cell structures (i.e., up to $20F^2$) are integrated, therefore limiting the array size. This is why the 1T-1R RRAM arrays are best suited for embedded applications where the latencies dominate the trade-off between speed and storage density. Several arrays realizations have been proposed in literature using different transistor technologies (i.e., full-CMOS or BiCMOS) and materials (HfO_2, Cu_xO, etc.) starting from few kbits structures up to a recently demonstrated 16 Gbits array [23, 46, 47].

To achieve higher densities, the transistor element can be replaced either by a diode (i.e., 1D-1R) or by any other kind of non-linear switching elements (i.e., 1S-1R) [10]. The diode selector allows achieving, on top of RRAM cells, a two terminal selector (smaller cell size are possible compared to 1T-1R), but it's valid only for RRAM cells that display a unipolar switching characteristics that is typical for non-valence chemical memories [10]. To adapt this concept to traditional RRAM cells (i.e., bipolar switching) several selectors can be adopted: symmetric diodes

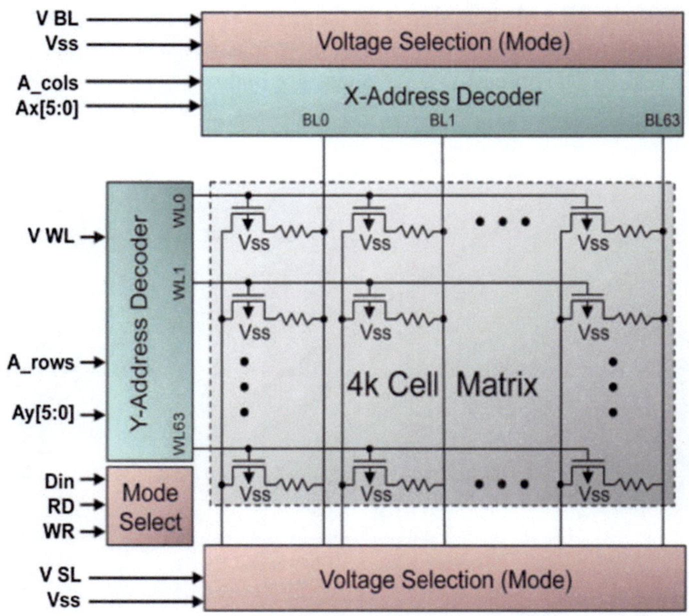

Fig. 5.18 Block diagram of a 4 kbits 1T-1R RRAM array [46]

or volatile switching selectors. The parameters of these selectors can be compared in [49].

5.4.3 3D RRAM Array Options: 1T-nR and VRRAM

To further increase the density of RRAM arrays several researchers started to consider the frontier of the third integration dimension, giving rise to the 3D RRAM array concepts. Two main process options exist for 3D RRAM: the pseudo 3D also indicated as 1T-nR or stackable 2D, and the vertical RRAM (VRRAM).

Concerning the 1T-nR approach, this is obtained by stacking multiple 2D RRAM planes, each one selected by a proper decoding structure [10, 49]. The 1T-nR arrays often utilize cross-point core architectures for higher density and one transistor that drives n RRAM devices. Unfortunately, multiple leakage current paths exist as shown in Fig. 5.19 even if a good transistor is fabricated, therefore requesting the use of the selectors presented in the previous sections.

A solution for leakage path suppression is to use a selector with a very high non-linearity in order to have a *high selectivity ratio* (HSR) of the cells, maintaining the scalability of the memory cells. The selectivity is the feature of the device that will activate the cell based on the potential across the two terminal RRAM. An example is the *Field Assisted Superlinear Threshold* (FAST) device presented in [49] (see Fig. 5.20). In any case, the selector must ensure a number of parameters that are compatible with 3D integration like: the speed of operation (i.e., a selector must switch in the same RRAM switching time range that sometimes could approach tens of nanoseconds), low power consumption (i.e., low leakage), and high compatibility with the CMOS BEOL process (i.e., temperature budget).

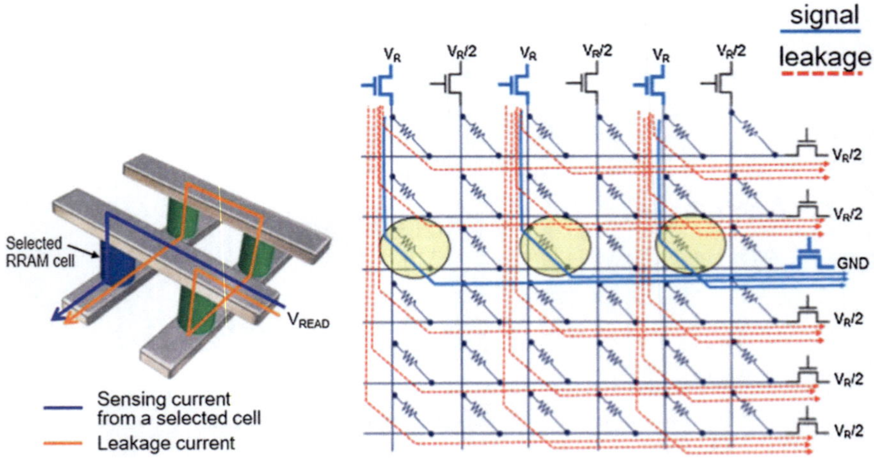

Fig. 5.19 Signal and leakage path in 1T-nR RRAM architecture [49]

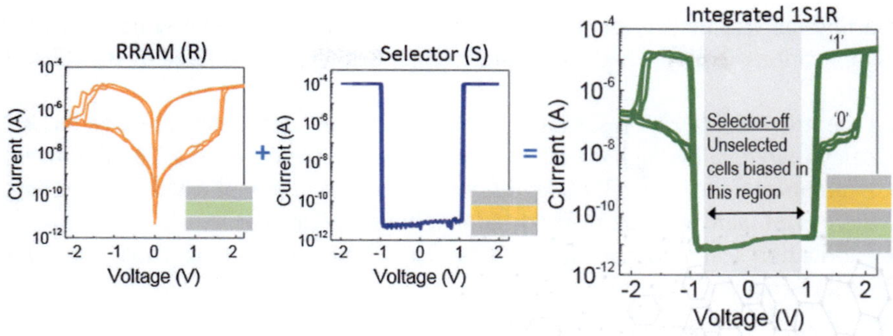

Fig. 5.20 RRAM integration with the FAST selector for pseudo-3D arrays [49]

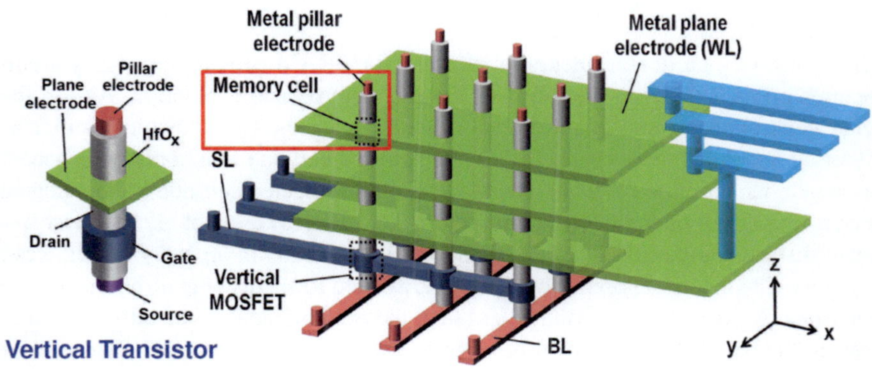

Fig. 5.21 Structure of a VRRAM integrating an HfO_x switching oxide [50]

A real paradigm shift could come from the VRRAM structures presented in [50]. Unfortunately, even if the promise of large density memories is kept, there are still a number of issues that limit the memory endurance up to 600 Set/Reset cycles. The switching speed of those memories is in the range of tens of nanoseconds. Figure 5.21 shows and example of a VRRAM that exploits vertical transistors for the cell selection stage in the array and the integration of the MIM switching element in a pillar that passes through a metal electrode plate.

The schemes applied for Set/Reset and Read operation in VRRAM are the same of 1T-nR RRAM, where the select transistor acts both as current limiter and as cell enabler in the array.

Summarizing, the challenges for 1T-nR and VRRAM widespread manufacturing and adoption are the following [50]:

- Sneak path leakage: the current sneak paths causing disturbances must be removed also to reduce unwanted power consumption;
- Selection device: the selector (i.e., transistor or any other selecting element) must ensure a sufficient HRS/LRS ratio to discriminate between logical states, feature

a high current density to support high compliance currents, be scalable for 3D applications, exhibit a high endurance compatible with that offered by the RRAM cell;
- Line resistance, Plane resistance, and Parasitic capacitance: to minimize the RC delays and to allow high operation speed those parameters need to be minimized;
- Interconnections: since 3D integration could result in complicated wire routing, there is an increased circuit overhead that must be reduced;
- Fabrication technologies: the pillar structures require etching capabilities for CMOS-friendly material.

5.5 Typical Disturbs in RRAM Technology

The choice of a particular integration scheme for RRAM cells in large array architectures poses a challenge in terms of disturb analysis and evaluation. As for any other generation of non-volatile memory technology, the term "disturb" indicates either an unwanted state change of an unaddressed cell due to a continuous access on neighbors addressed cells or a state change of a selected cell due to its repeated access like a continuous Read or data re-write. In RRAM technology the most frequent disturb sources are in selected and unselected cells in an array during Read operations. In the most popular RRAM cells, Set and Reset occur at different voltage polarities. A positive Read voltage less than the Set voltage is frequently chosen to prevent the Set (LRS)-state disturb, whereas the Reset (HRS)-state disturb must be carefully engineered [51].

On crossbar-based arrays, which is one of the potential array integration topologies offered by RRAM technology, unselected wordlines and bitlines can be grounded or biased with a $V_{dd}/3$ or $V_{dd}/2$ scheme, as presented in the earlier sections. In order to evaluate the impact on unselected cells during Set/Reset operations in the worst-case condition, the $V_{dd}/2$ biasing effect on Reset and Set wordlines has been evaluated on different sized 1T-1R RRAM arrays mimicking the access modes of a cross-point array (this is possible because in 1T-1R arrays all the cells can be accessed individually). $10^6 V_{Reset}/2$ pulses have been applied on wordlines in Set state (LRS), while $10^6 V_{Set}/2$ pulses have been applied on wordlines in Reset state (HRS), where V_{Reset} and V_{Set} are the average Reset and Set switching voltages of the RRAM cells in the array, respectively. The disturb effect has been evaluated on both fresh and cycled devices, after 10 k Set/Reset cycling operations [46]. Figure 5.22 shows the average Set/Reset state read currents and their standard deviation measured during 10^6 stress pulses on $0.6\,\mu m^2$ (Fig. 5.22a) and $1\,\mu m^2$ (Fig. 5.22b) devices, for both fresh and cycled arrays. The dielectric material degradation in the MIM stack makes Reset and Set switching less effective, reducing the stress sensibility as well. The average current variation observed during stress is depicted in Fig. 5.23. The $V_{dd}/2$ stress caused a higher read current shift on fresh devices, for both $0.6\,\mu m^2$ and $1\,\mu m^2$ 1T-1R RRAM devices.

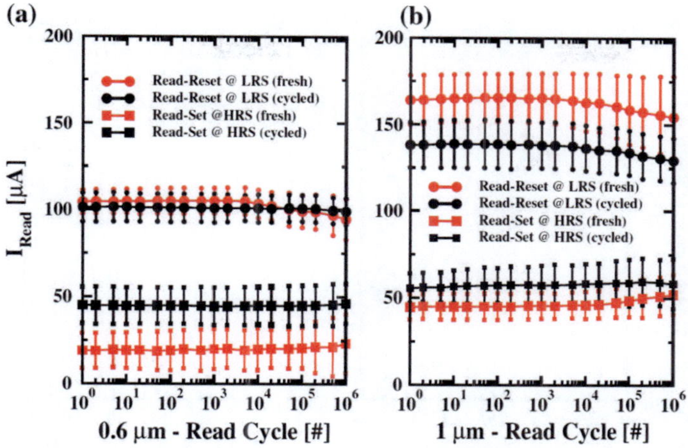

Fig. 5.22 $V_{dd}/2$ pulse stress effect measured on 0.6 μm^2 (**a**) and 1 μm^2 1T-1R RRAM arrays (**b**), in both set (LRS) and reset (HRS) condition [46]

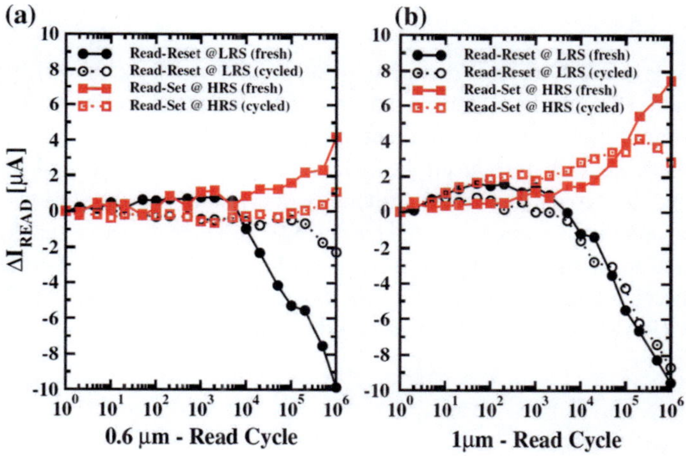

Fig. 5.23 Average read current variation measured during $V_{dd}/2$ pulse stress measured on 0.6 μm^2 (**a**) and 1 μm^2 1T-1R RRAM devices (**b**) [46]

Ideally, considering device and circuit design margins, read resistance variation should be less than 10%. Error Correction Codes can also assist in recovery from less frequent, larger resistance fluctuations, but the occurrence of the resistance variation should be less than 1% for effective data integrity. The Read Error Rate, calculated as the fraction of cells showing a resistance variation higher than 10% during $V_{dd}/2$ stress is depicted in Fig. 5.24. Fresh devices show a higher error rate than cycled devices (after 10 k Set/Reset cycles). Although the average read current variation is higher for large RRAM cell area, the error rate is lower with respect to that of smaller devices because of a higher average Set and Reset currents that render the

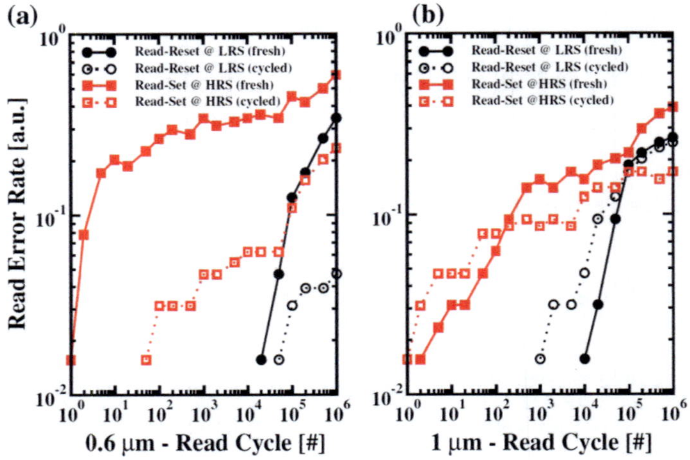

Fig. 5.24 Read Error Rate calculated on 0.6 μm^2 (**a**) and 1 μm^2 1T-1R RRAM devices (**b**). *Full and dotted lines* refers to fresh and cycled devices, respectively [46]

fluctuations less effective. Read disturb with Set polarity (LRS) stress on fresh devices in Reset state (HRS) is the operation that shows the highest read error rate due to the conformation of the conductive filament in the cells [46, 51].

The present chapter introduced the RRAM technology from several standpoints by exposing the typical operation principles, the architectural solutions for their integration, and the limitations caused by intrinsic and extrinsic aspects. To evaluate the exploitation of the RRAM for storage application like Solid State Drives we would redirect the reader to Chap. 6, where simulation aspects will be devoted to devise the RRAM features for high bandwidth and low latency storage.

References

1. J. Suhonen et al., "Low-Power Wireless Sensor Networks: Protocols, Services and Applications," Springer, 2012.
2. B. de Salvo, "Silicon Non-Volatile Memories: Paths of Innovation," Hoboken, Wiley-ISTE, 2009.
3. Y. Zhang, "Future Wireless Networks and Information Systems," Springer, 2012.
4. L. O. Chua, "Resistance switching memories are memristors," Applied Physics A, vol. 102, no. 4, pp. 765–783, 2011.
5. R. Waser and M. Aono, "Nanoionics-based resistive switching memories," Nature Mater., vol. 6, pp. 833–840, 2007.
6. J. Lee et al., "Effect of ZrO_X/HfO_X bilayer structure on switching uniformity and reliability in nonvolatile memory applications," Appl. Phys. Lett., vol. 97, no. 17, p. 172–105, Oct. 2010.
7. Ch. Walczyk et al., "Impact of Temperature on the Resistive Switching Behavior of Embedded HfO_2-Based RRAM Devices," IEEE Trans. Electron Devices, vol. 58, no. 9, pp. 3124–3131, 2011.

8. D. Walczyk et al., "Resistive switching behavior in TiN/HfO2/Ti/TiN devices," International semiconductor conference dresden-grenoble (ISCDG), 2012, pp. 143–146.
9. P. Lorenzi et al., "Forming kinetics in HfO2-based RRAM cells," IEEE Trans. Electron. Devices, vol. 60, no. 1, pp. 438–443, 2013.
10. N. Raghavan et al., "Statistical insight into controlled forming and forming free stacks for HfOx RRAM", Microelectron. Eng., vol. 109, pp. 177–181, 2013.
11. D. Wouters, "Resistive switching materials and devices for future memory applications," Tutorial IEEE-SISC, Dec. 2012.
12. T. Ninomiya et al., "Conductive filament scaling of TaOx bipolar ReRAM for improving data retention under low operation current," IEEE Trans. Electron Devices, vol. 60, no. 4, pp. 1384–1389, 2013.
13. H-T. Liu et al., "Effect of pulse and dc formation on the performance of one-transistor and one-resistor resistance random access memory devices," Chin. Phys. Lett., vol. 32, no. 2, pp. 1–3, 2015.
14. C. Zambelli et al., "Statistical analysis of resistive switching characteristics in ReRAM test arrays," IEEE International Conference on Microelectronics Test Structures (ICMTS), pp. 27–31, 2014.
15. A. Grossi et al., "Electrical characterization and modeling of pulse-based forming techniques in RRAM arrays," Solid-State Electronics, vol. 115, Part A, pp. 17–25, 2016.
16. B. Govoreanu et al., "10×10nm2 Hf/HfOx crossbar resistive RAM with excellent performance, reliability and low-energy operation," IEEE International Electron Devices Meeting (IEDM), 2011, pp. 31.6.1–31.6.4.
17. A. Grossi et al., "Performance and reliability comparison of 1T-1R RRAM arrays with amorphous and polycrystalline HfO2," Joint International EUROSOI Workshop and International Conference on Ultimate Integration on Silicon (EUROSOI-ULIS), 2016, pp. 80–83.
18. E.A. Miranda et al., "Model for the Resistive Switching Effect in HfO2 MIM Structures Based on the Transmission Properties of Narrow Constrictions," IEEE Electron Device Letters, vol. 31, no. 6, pp. 609–611, 2010.
19. S. Ambrogio et al., "Statistical Fluctuations in HfOx Resistive-Switching Memory: Part I - Set/Reset Variability," IEEE Trans. on Electron Devices, vol. 61, no. 8, pp. 2912–2919, 2014.
20. C. Zambelli et al., "Electrical characterization of read window in reram arrays under different SET/RESET cycling conditions," IEEE International Memory Workshop (IMW), 2014, pp. 1–4.
21. G. Wang et al., "Impact of stress time of program operation on the endurance performance," IEEE International Conference on Solid-State and Integrated Circuit Technology (ICSICT), 2014, pp. 1–3.
22. Y.Y. Chen et al., "Balancing SET/RESET Pulse for 10^{10} Endurance in HfO_2/Hf 1T1R Bipolar RRAM," in IEEE Trans. on Electron Devices, vol. 59, no. 12, pp. 3243–3249, 2012.
23. S.-S. Sheu et al., "A 4Mb embedded SLC resistive-RAM macro with 7.2 ns read-write random-access time and 160 ns MLC-access capability," Proc. ISSCC, 2011, pp. 200–202.
24. Y.S. Chen et al., "Highly Scalable Hafnium Oxide Memory with Improvements of Resistive Distribution and Read Disturb Immunity," IEEE International Electron Devices Meeting (IEDM), 2009, pp. 105–108.
25. T.-Y. Liu et al., "A 130.7 mm2 2-Layer 32 Gb ReRAM Memory Device in 24 nm Technology," Proc. ISSCC, 2013, pp. 210–212.
26. A. Grossi et al., "Impact of Intercell and Intracell Variability on Forming and Switching Parameters in RRAM Arrays," in IEEE Trans. on Electron Devices, vol. 62, no. 8, pp. 2502–2509, 2015.
27. A. Fantini et al., "Intrinsic switching variability in HfO2 RRAM," IEEE International Memory Workshop (IMW), 2013, pp. 30–33.
28. D. Ielmini and R. Waser, "Resistive Switching: From Fundamentals of Nanoionic Redox Processes to Memristive Device Applications", Wiley, 2016.
29. B. Jiao et al., "Resistive switching variability study on 1T1R AlOx/WOx-based RRAM array," IEEE International Conference of Electron Devices and Solid-State Circuits (EDSSC), 2013, pp. 1–2.

30. A. Chen and M. Lin, "Variability of resistive switching memories and its impact on crossbar array performance," in IEEE International Reliability Physics Symposium (IRPS), 2011, pp. MY.7.1–MY.7.4.
31. G. Piccolboni et al., "Investigation of HfO2/Ti based vertical RRAM - Performances and variability," Non-Volatile Memory Technology Symposium (NVMTS), 2014, pp. 1–5.
32. Y.Y. Chen et al., "Understanding of the Endurance Failure in Scaled HfO2-based 1T1R RRAM through Vacancy Mobility Degradation", IEDM Tech. Dig., 2012, pp.482–485.
33. B. Chen et al., "Physical Mechanisms of Endurance Degradation in TMO-RRAM", IEDM Tech. Dig., 2011, pp.283–286.
34. P. Huang et al., "Analytic model of endurance degradation and its practical applications for operation scheme optimization in metal oxide based RRAM," IEEE International Electron Devices Meeting (IEDM), 2013, pp. 22.5.1–22.5.4.
35. K. Higuchi et al., "Investigation of Verify-Programming Methods to Achieve 10 Million Cycles for 50nm HfO2 ReRAM," IEEE International Memory Workshop (IMW), 2012, pp. 1–4.
36. H. Yamazawa et al., "50 nm AlxOy ReRAM array retention characteristics before and after endurance," Silicon Nanoelectronics Workshop (SNW), 2014, pp. 1–2.
37. Y.Y. Chen et al., "Improvement of data retention in HfO2/Hf 1T1R RRAM cell under low operating current," IEEE International Electron Devices Meeting (IEDM), 2013, pp. 10.1.1–10.1.4.
38. S. Yu et al., "A Monte Carlo study of the low resistance state retention of HfOx based resistive switching memory," Applied Physics Letters, vol. 100, 043507, 2012.
39. D. Ielmini et al., "Size-Dependent Retention Time in NiO-Based Resistive-Switching Memories," in IEEE Electron Device Letters, vol. 31, no. 4, pp. 353–355, 2010.
40. F.M. Puglisi et al., "Instability of HfO2 RRAM devices: Comparing RTN and cycling variability," IEEE International Reliability Physics Symposium (IRPS), 2014, pp. MY.5.1–MY.5.5.
41. A. Grossi et al., "Relationship among Current Fluctuations during Forming, Cell-To-Cell Variability and Reliability in RRAM Arrays," IEEE International Memory Workshop (IMW), 2015, pp. 1–4.
42. R. Degraeve et al., "Quantitative model for post-program instabilities in filamentary RRAM," presented at IEEE International Reliability Physics Symposium (IRPS), 2016.
43. J. Liang and H.S.P. Wong, "Cross-Point Memory Array Without Cell Selectors—Device Characteristics and Data Storage Pattern Dependencies," in IEEE Trans. on Electron Devices, vol. 57, no. 10, pp. 2531–2538, 2010.
44. A. Sawa, "Resistive switching in transition metal oxides," in Materials Today, vol. 11, no. 6, pp. 28–36, 2008.
45. X. Xue et al., "A 0.13 μm 8 Mb Logic-Based Cu_xSi_yO ReRAM With Self-Adaptive Operation for Yield Enhancement and Power Reduction," in IEEE Journal of Solid-State Circuits, vol. 48, no. 5, pp. 1315–1322, 2013.
46. C. Zambelli et al., "RRAM Reliability/Performance Characterization through Array Architectures Investigations," IEEE Computer Society Annual Symposium on VLSI (ISVLSI), 2015, pp. 327–332.
47. R. Fackenthal et al., "A 16Gb ReRAM with 200MB/s write and 1GB/s read in 27nm technology," IEEE International Solid-State Circuits Conference Digest of Technical Papers (ISSCC), 2014, pp. 338–339.
48. H. Nazarian, "Breakthrough 3D RRAM Technology for super-dense, low latency, low power data storage system", Flash Memory Summit, 2014.
49. S.H. Jo, "Recent Progress in RRAM Materials and Devices", SEMICON Korea, 2015.
50. H.H.-Y. Chen et al., "3D Vertical RRAM," Flash Memory Summit, 2013.
51. W.-C. Luo et al., "Rapid Prediction of RRAM RESET-State Disturb by Ramped Voltage Stress", in IEEE Electron Device Letters, vol. 33, no. 4, 2012.

Chapter 6
Simulations of RRAM-Based SSDs

Lorenzo Zuolo, Cristian Zambelli, Rino Micheloni and Piero Olivo

Abstract In this chapter SSDExplorer, a fine-grained SSD simulator, is used to evaluate the possible impact of emerging non-volatile memories, such as Resistive RAM (RRAM), on future SSD architectures. Does it make sense to fully replace NANDs with one of the emerging memories? What's the benefit? Is it better to develop a hybrid drive, where a RRAM is used as cache? Most of the new memories are still in the development phase, well before a real mass production; as a matter of fact, simulations are the only way to answer the above mentioned questions and figure out where a new non-volatile technology can really help, in terms of both performances and cost saving.

SSDs are the most effective solution for both consumer applications and large enterprise environments when high performance storage devices are required [1]. To cope with the increasing request of data storage, especially for large computing facilities, there is a call for a continuous expansion of the bit density in the SSD storage medium, namely the NAND Flash. This is generally achieved through either a technology shrink or a multi-bit per cell storage or both; in all cases, it implies a significant degradation of memory speed and reliability, thus impacting the main figures of merit of an SSD (i.e., latency and bandwidth) [2].

Resistive RAM (RRAM) is perceived by the storage community as a reliable alternative to NAND Flash in SSDs for low latency applications [3]. These emerging memories are non-volatile as NAND flash, but with a lower read/write latency and a higher reliability. However, the relatively small storage capacity of RRAM memories integrated so far [4, 5] has limited their usage to specific applications such as saving critical data during power loss events or as a cache memory for fast data manipulation, like in the hybrid system described in [6]. In this case, RRAMs are combined with

L. Zuolo (✉) · C. Zambelli · P. Olivo
Dipartimento di Ingegneria, Università degli Studi di Ferrara, via G. Saragat, 1, 44122 Ferrara, Italy
e-mail: lorenzo.zuolo@unife.it

R. Micheloni
Performance Storage Business Unit, Microsemi Corporation, Vimercate, Italy

R. Micheloni (ed.), *Solid-State-Drives (SSDs) Modeling*, Springer Series in Advanced Microelectronics 58,
DOI 10.1007/978-3-319-51735-3_6

NAND flash memories to minimize latency and to improve both the bandwidth and the reliability of the drive.

To increase the density of RRAM memory arrays, several researchers started to consider advanced 3D architectures. Among them, the 1T-nR approach seems the easiest to integrate by stacking multiple RRAM planes, each one selected by a proper decoding structure [7]. The 1T-nR arrays often utilize cross-point core architectures for higher density and one transistor that drives n RRAM devices. Those arrays are forecasted to be fully compatible with the state-of-the-art NAND Flash interface [8, 9], paving the way to innovative "All-RRAM" SSD's architectures. In these systems, NAND flash memories are completely replaced by RRAM devices offering a highly reliable and extremely faster storage medium.

In this chapter, a thorough design space exploration of a 512 GB All-RRAM SSD architecture is performed, with particular attention to architectural bottlenecks and inefficiencies, by using the SSDExplorer co-simulator [10]. We assumed a full compatibility of RRAM chips with typical NAND flash interfaces [11, 12], and hence a state-of-the-art SSD controller is embodied in the simulation environment. In light of these considerations, we leverage both the internal page architecture of a 1T-nR RRAM chip [8] and the SSD's firmware to find the optimal configuration, thus enabling the adoption of the RRAM technology in high performance SSD applications. Collected results show that, in standard working condition (i.e., when 4 kB transactions are issued by the host system), All-RRAM SSDs are able to show extremely low latency only if a proper management of the operations is adopted.

6.1 All-RRAM SSD Architecture

The RRAM chip considered in the simulated SSD architecture is a configurable 16 planes 32 Gbits memory module with a 8 bit ONFI 2.0/Toggle Mode interface, capable of 200 MB/s [11, 12] (see Fig. 6.1). Each plane is a 2 Gbit RRAM array with a page size of 256 B [8]. The RRAM chip features an internal memory controller that can work either with a native addressing mode (i.e. 256 byte-wide page) or in a multi-plane emulated addressing mode, which allows accessing from 512 B up to 4 kB within a single operation. A read operation takes 1 μs per page. The main array characteristics of this technology are summarized in Table 6.1.

The simulated SSD configuration (sketched in Fig. 6.2) is a 512 GB drive made of 16 channels; each channel is populated with 8 RRAM targets. SSD controller, Error Correction Code modules and DRAM buffers are included to keep the compatibility with state-of-the-art NAND-Flash based SSDs [10]. The drive host interface is a PCI-Express Gen2 with 8 lanes adopting the NVM-Express protocols [13], which is typical in enterprise-class SSDs. Different 100% random read workloads were used to test the 2 addressing modes described above, by aligning the logical block address of the drive with the effective RRAM page size. Write workloads are not described

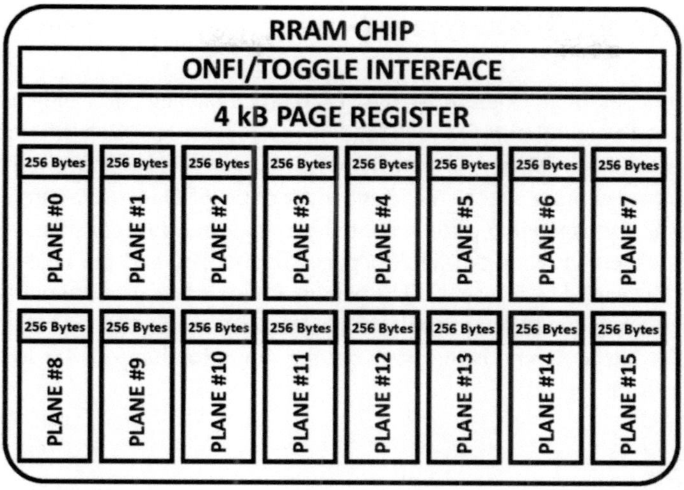

Fig. 6.1 32 Gbit RRAM memory module architecture

Table 6.1 Main characteristics of the simulated RRAM devices

Chip parameter	Configuration
IO-bus interface	ONFI 2.0/Toggle mode
IO-bus speed	200 MB/s
Native page size	256 B
Emulated page size	512–1024–4096 B
t_{READ} per page	1 μs

Fig. 6.2 Block diagram of the simulated All-RRAM 512 GB SSD architecture

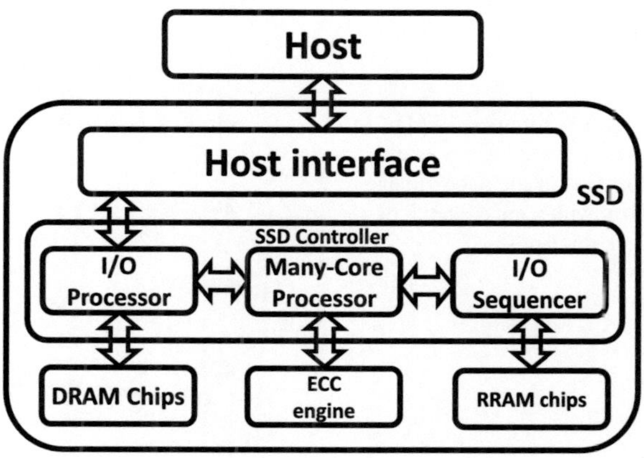

in this chapter since the target drive architecture makes use of DRAM buffers where write operations are cached; therefore they do not represent a significant threat for the latency and bandwidth figures.

6.1.1 Page Size Versus Queue Depth

A statistical assessment of the All-RRAM SSD read latency and bandwidth figures was performed by simulating 500,000 random read operations, with different RRAM page sizes and queue depths. As shown in Figs. 6.3 and 6.4, by setting a fixed queue depth of 16 read commands, the read bandwidth of the SSD increases proportionally with the memory page size, while the latency remains almost constant. A different behavior is observed when the RRAM page size is fixed and the read commands queue depth is varied from 1 to 32. The bandwidth increases proportionally with the queue depth up to a saturation level, which depends on the RRAM page size (see Fig. 6.5). For latency, trend is the same but the saturation happens for lower queue depth values (see Fig. 6.6).

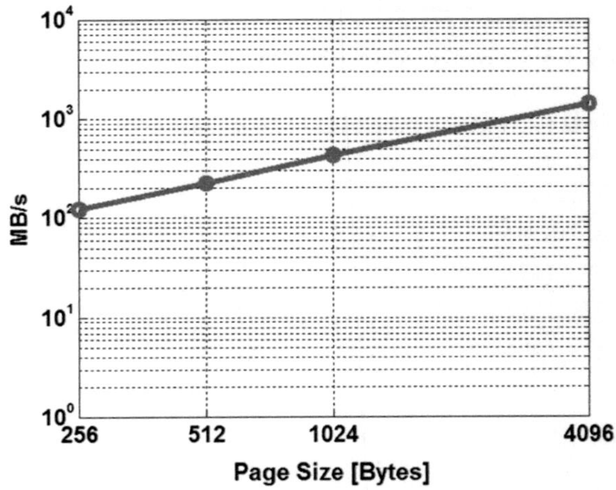

Fig. 6.3 Simulated SSD's average read bandwidth as a function of the RRAM page size, with a queue depth of 16 commands

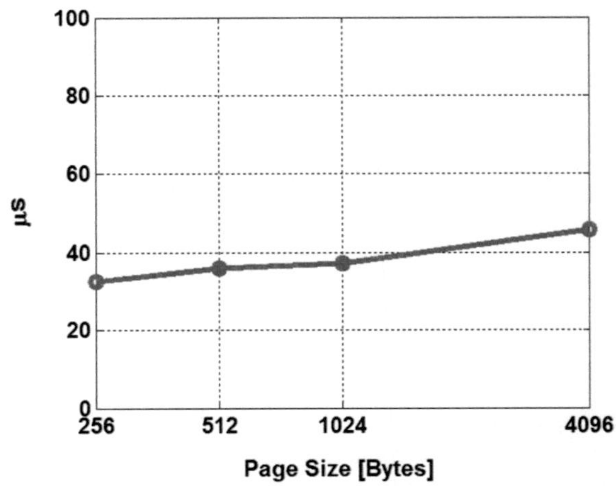

Fig. 6.4 Simulated SSD's average read latency as a function of the RRAM page size, with a queue depth of 16 commands

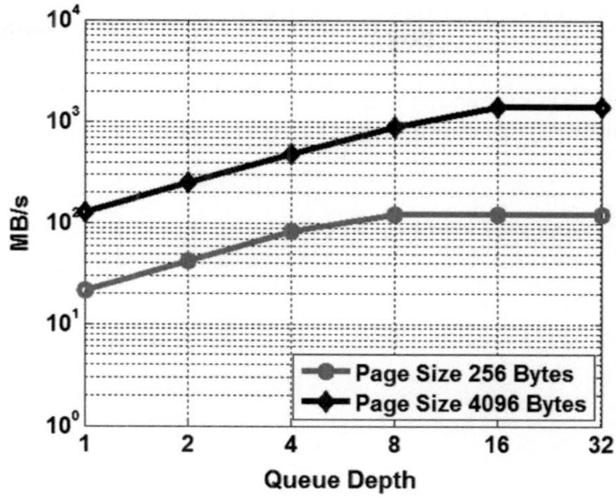

Fig. 6.5 Simulated SSD's average read bandwidth as a function of the host interface queue depth. Native 256 B and 4 kB multi-plane RRAM addressing modes are considered

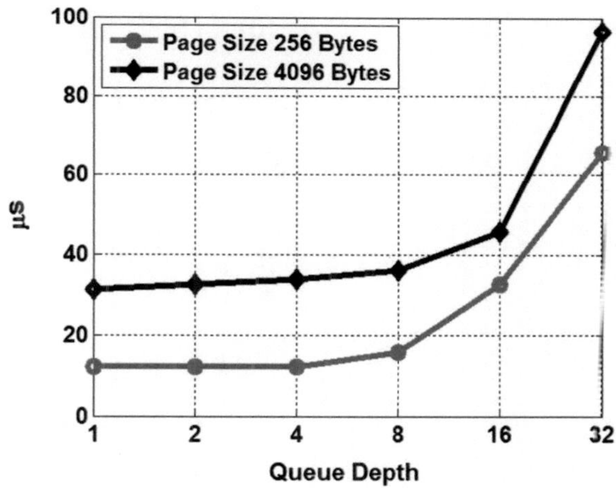

Fig. 6.6 Simulated SSD's average read latency as a function of the host interface queue depth. Native 256 B and 4 kB multi-plane RRAM addressing modes are considered

In SSD architectures, especially those for enterprise environments, it is extremely important to analyze the *Quality of Service* (QoS), with special focus on read. A slower QoS of the read operation corresponds to a longer response time of the drive when the host wants to read data for subsequent manipulation [14]. The *Cumulative Distribution Function* (CDF) and the *Probability Density Function* (PDF) of the latency are very useful tools for this kind of analyses. Figure 6.7 shows CDF and PDF for the All-RRAM SSD. When user transactions match the RRAM chip page size (i.e. 256 B), and only one operation is served at a time, latency gets extremely low, in the range of tens of microseconds. The longest read response time (i.e., 99.99 percentile of the CDF) is around 16 μs, which is well below the few hundreds of microseconds offered by NAND Flash-based SSDs. However, such queue depth and RRAM page size do not reflect the workload conditions of the state-of-the-art host platforms and file-systems, which are designed to issue multiple read operations with

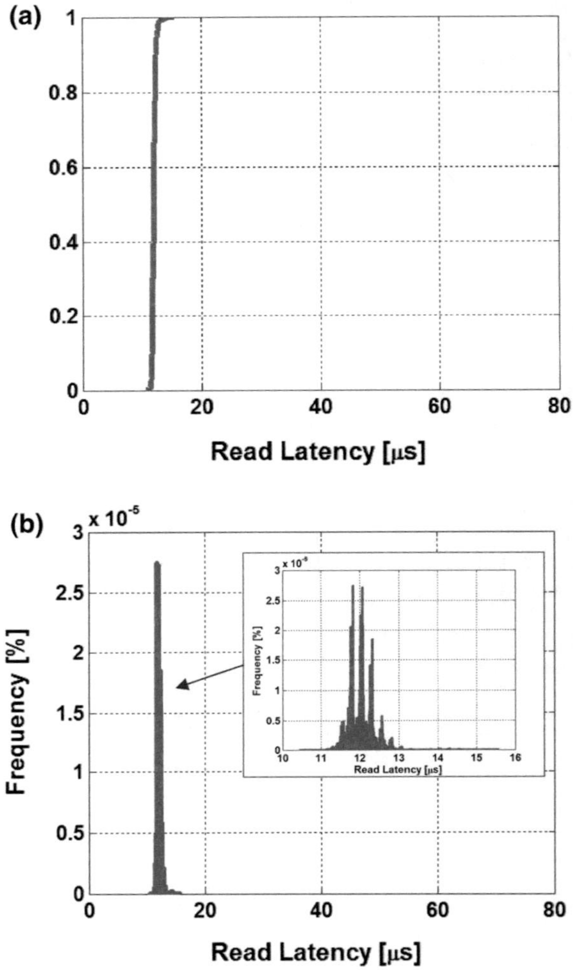

Fig. 6.7 CDF (**a**) and PDF (**b**) of the simulated SSD's read latency with a queue depth of 1 command and a native 256 B RRAM page size

a fixed payload of 4 kB. Looking at Fig. 6.8, when the host interface queue depth is fixed to 32 commands and user operations match the native 256 B RRAM addressing mode, the median latency rapidly increases up to 66 μs. Eventually, with a 4 kB page All-RRAM SSD, read response times become very similar to those of a simulated 1x-nm MLC NAND Flash-based SSD (Fig. 6.9).

In order to explain the above mentioned results, we observed the RRAM I/O bus interface utilization and the percentage of active RRAM dies, as a function of the RRAM page size and drive's commands queue depth. When the payload of read transactions increases and the queue depth is large enough to serve multiple commands, more data have to be transferred from the memories to the SSD controller. This condition yields to a massive overhead in terms of data transfers, thus impacting the percentage of the I/O memory bus usage. This metric rapidly grows up, reaching 48%

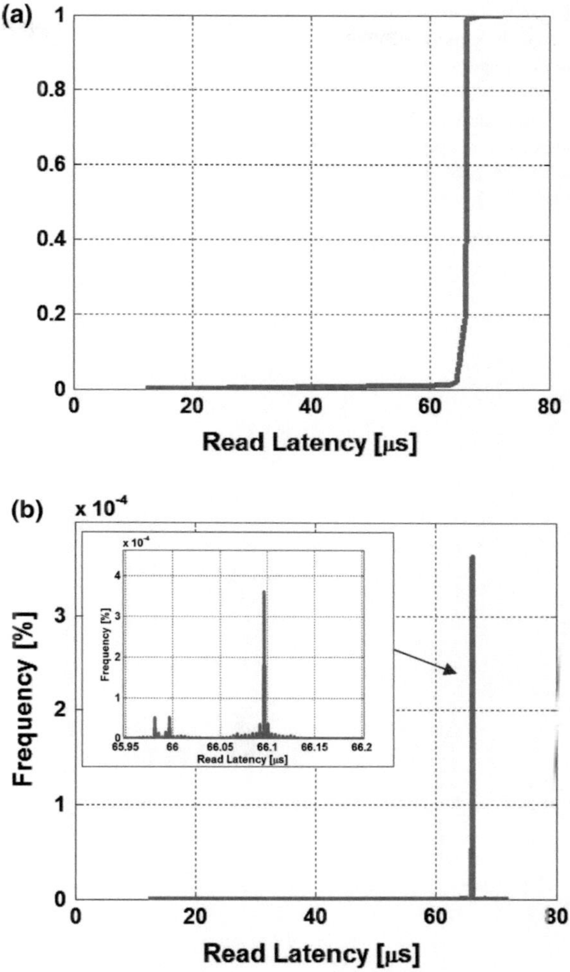

Fig. 6.8 CDF (**a**) and PDF (**b**) of the simulated SSD's read latency with a queue depth of 32 commands and a native 256 B RRAM page size

when 4 kB transactions are served with a queue depth of 16, as shown in Figs. 6.10 and 6.11. As a consequence, the overall SSD latency is impacted and the performance advantages of RRAMs partially vanish. Another important consideration can be made by observing the average percentage of active RRAM memories under a 100% random read workload. With reference to Figs. 6.12 and 6.13, even considering 4 kB transactions, this percentage remains far below 10%. These results clearly denote a high under-utilization of SSD resources. In fact, as previously described, the analyzed All-RRAM SSD is based on a controller designed for NAND flash memories. This basic approach is cost-effective but, on the other hand, it does not permit to properly use the underlying storage medium, which is completely different from NAND.

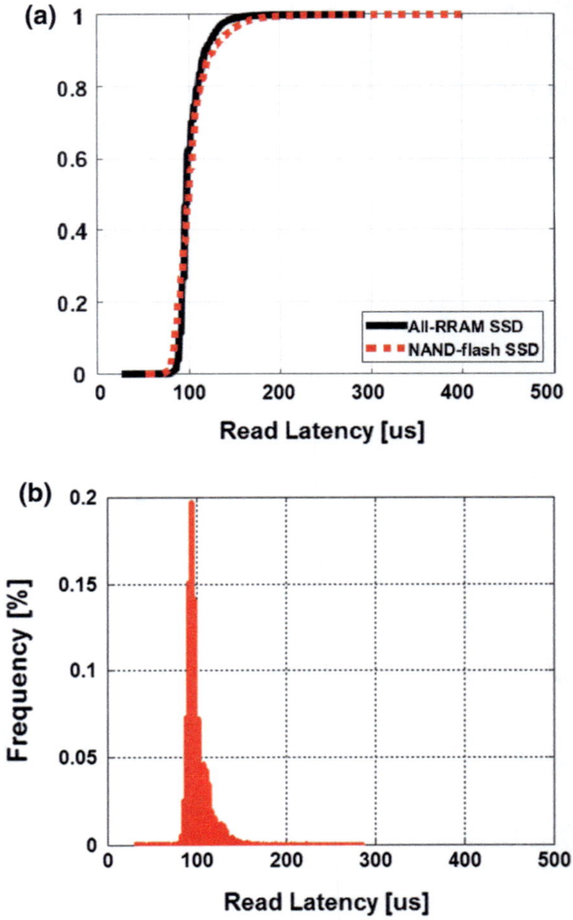

Fig. 6.9 CDF (**a**) and PDF (**b**) of the simulated SSD's read latency with a queue depth of 32 commands and the emulated 4 kB RRAM page size. A comparison with a state-of-the-art NAND Flash SSD is provided

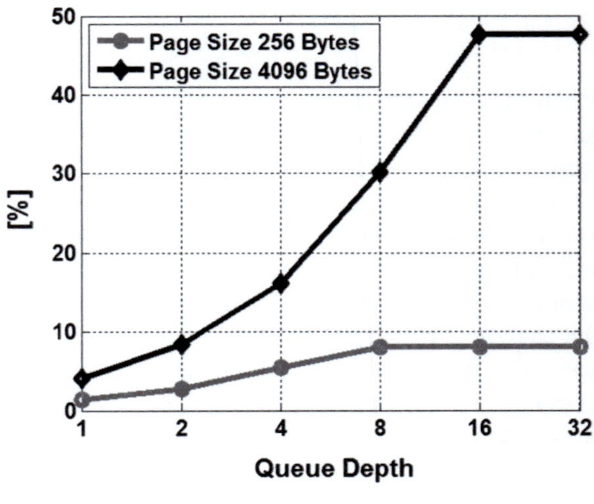

Fig. 6.10 Average RRAM I/O bus interface usage as a function of the host interface queue depth. Native 256 B and 4 kB multi-plane RRAM addressing modes are considered

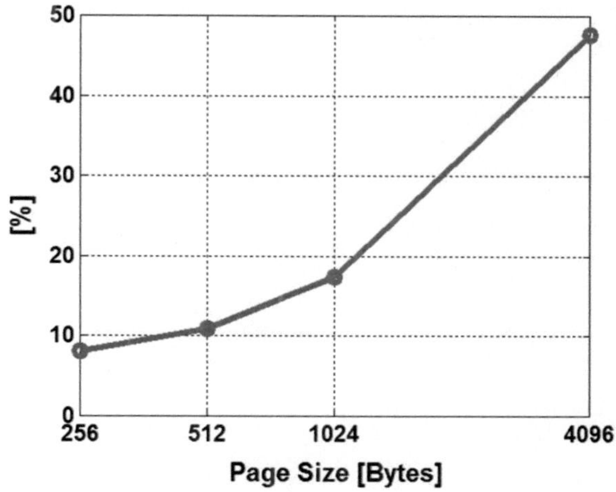

Fig. 6.11 Average RRAM I/O bus interface usage as a function of the RRAM page size when a queue depth of 16 commands is fixed

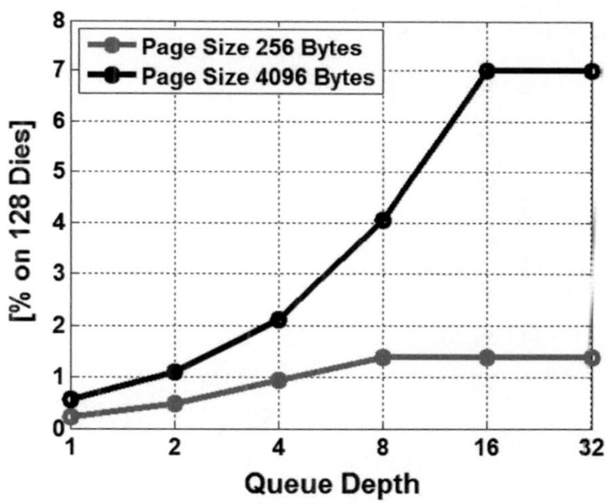

Fig. 6.12 Percentage of active RRAM dies as a function of the host interface queue depth. Native 256 B and 4 kB multi-plane RRAM addressing modes are considered

6.1.2 Design Space Exploration of All-RRAM SSDs

Figures 6.14 and 6.15 show a breakdown of the latency, considering a 200 MB/s DDR I/O bus frequency. It is clear that, compared to NAND flash memories, the I/O bus transfer time is the dominant factor when RRAMs are used.

Thanks to the advent of extremely fast storage media such as RRAMs, memory vendors are now investing to push the I/O frequency to 400 MHz and beyond [15].

In order to understand how next generation SSD controllers could improve performances of an All-RRAM SSD, a complete design space exploration was performed, considering a 800 MB/s I/O bus transfer rate and 5 different RRAM page sizes: 256 B, 512 B, 1 kB, 2 kB, and 4 kB. For these simulations, the RRAM characteristics summarized in Table 6.1 were kept unaltered, and only the IO-Bus speed was

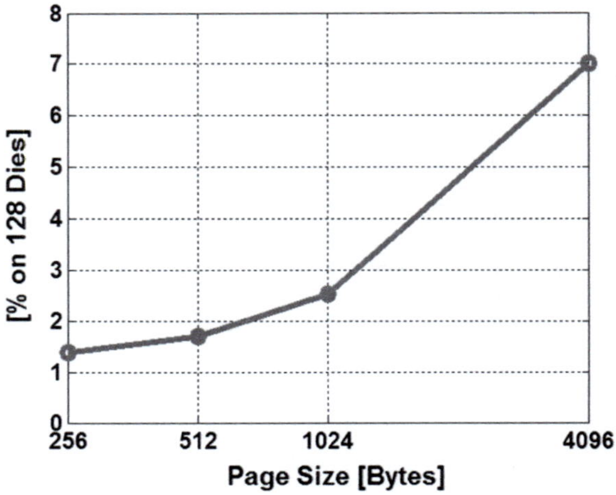

Fig. 6.13 Percentage of active RRAM dies as a function of the RRAM page size when a queue depth of 16 commands is fixed

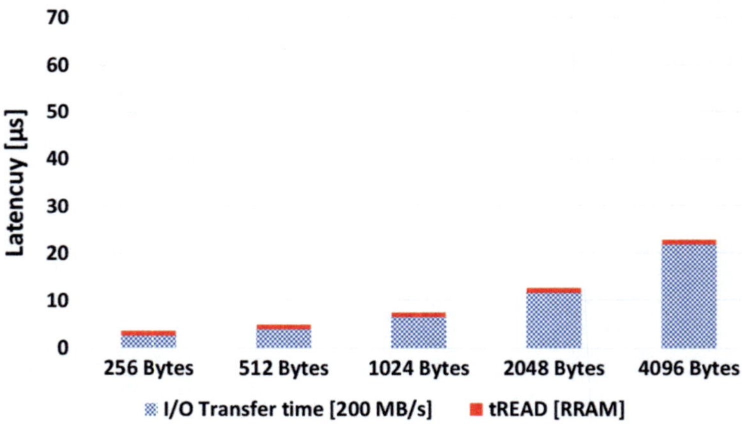

Fig. 6.14 Breakdown of the storage latency when a RRAM and a 200 MB/s I/O bus are used

increased to exploit the capabilities of the latest standard [15]. Drive bandwidth, average latency and Quality of Service (QoS) [14] were simulated for several page size configurations. The bandwidth is the average number of read commands completed in a second; the average latency is the average time elapsed between a read command submission and its completion; the QoS is computed as the 99.99 percentile of the SSD's latency distribution. To provide a complete performance exploration of the SSD's architecture, data were collected for different host *Queue Depths* (QD), ranging from 1 to 32 commands [13].

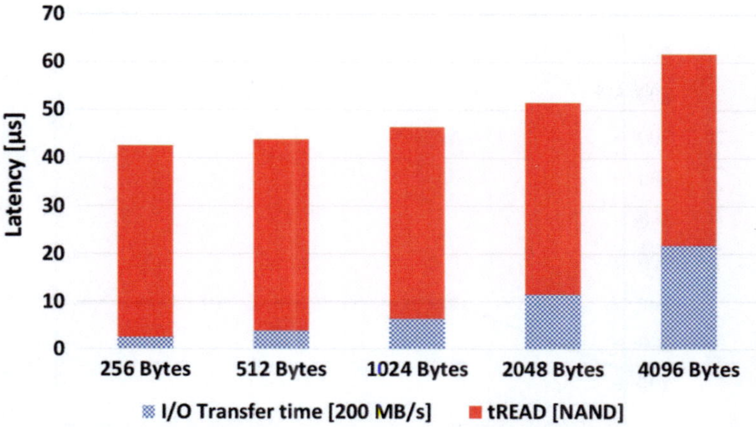

Fig. 6.15 Breakdown of the storage latency when a 1X-MLC NAND flash memory and a 200 MB/s I/O bus are used

6.1.2.1 NAND-like Mode: RRAM with 4 kB Page Size

This case study corresponds to the simple replacement of a NAND Flash memory with a RRAM chip in a user-transparent mode, and it will be used as a baseline for comparison. Therefore, as already presented in Sect. 6.1.1, in order to provide a full compatibility with NAND, the RRAM die must operate in 16-plane mode. Figures 6.16 and 6.17 (dashed lines) show average latency, QoS, and bandwidth

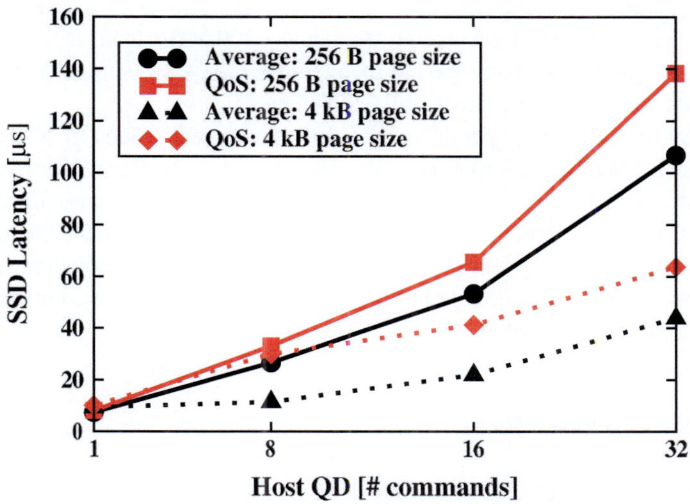

Fig. 6.16 Average latency and QoS of the simulated All-RRAM SSD with page sizes of 4 kB and 256 B

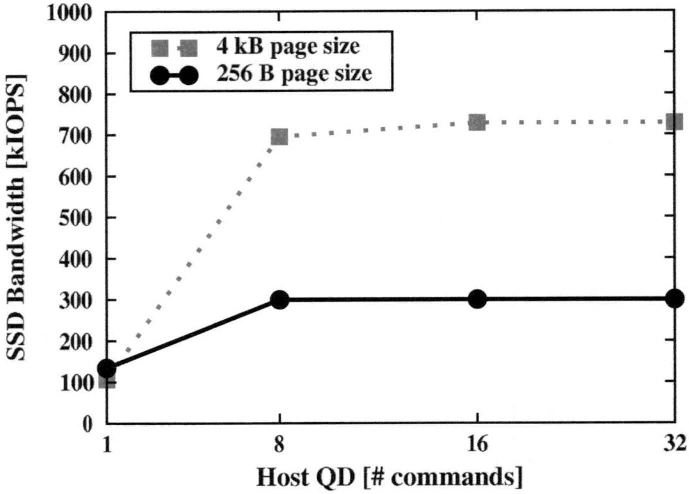

Fig. 6.17 Average bandwidth of the simulated All-RRAM SSD with page sizes of 4 kB and 256 B

increase with respect to QD. In particular, the bandwidth saturates for a QD equal to 8 commands, showing that the SSD controller has reached its maximum performance. As sketched in Fig. 6.14, the average read latency of the SSD's storage layer depends on two factors: memory t_{READ} and data transfer time from the memory to the SSD controller. As displayed in Fig. 6.16, at QD = 1 (i.e. one command issued at a time), average latency is 9.4 μs, which is almost 4 times shorter than the one observed in Fig. 6.6. This improvement is mainly due to the faster I/O bus transfer time (i.e., the 800 MB/s memory interface). In fact, in this case the transfer of a 4 kB page takes only 6 μs instead of the 21 μs taken by the legacy 200 MB/s interface. To be fair, it must be highlighted that, compared to the memory t_{READ} time, the I/O bus transfer contribution still dominates the overall SSD's latency.

Although the achieved latency is far below the typical values of NAND-based SSDs [8], RRAMs can be further optimized to reach even higher performances. For example, one area of improvement is the partitioning of the 4 kB transactions coming from the host into smaller chunks. The goal of this approach is to reduce both the data transfer time by selecting the right number of planes to be simultaneously accessed (i.e. optimal memory page size), and the number of internal read commands to be handled by the SSD firmware.

6.1.2.2 Single-Plane RRAM

The minimum read granularity allowed by RRAMs is a single plane 256 B page read operation, which could potentially reduce both the transfer time and the SSD latency. However, as shown in Fig. 6.18, since the host works with 4 kB transactions, it is necessary to split the host operations in 16 chunks of 256 B each. The firmware

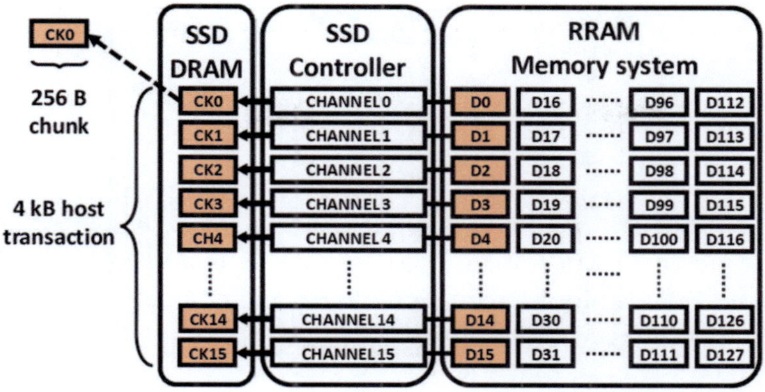

Fig. 6.18 A single 4 kB host transaction is split across multiple memory channels

running inside the SSD performs this operation: by using a 16 channels architecture and DRAM buffers, firmware reads in parallel all the addressed 256 B chunks, and rebuilds the 4 kB transaction before sending the data back to the host. Figures 6.16 and 6.17 (solid lines) show bandwidth, average latency and QoS achieved by the aforementioned approach. Looking at the results of the simulations performed with QD = 1, the straightforward conclusion would be that the 256 B page size reduces SSD latency, increases the bandwidth, and improves the QoS. However, for QD > 1 these considerations do not hold true anymore, since the number of operations internally handled by the SSD controller increases by a factor 16, leading to a saturation of its processing capabilities. This turns into a dramatic performance degradation, also considering that all data chunks must be temporarily stored inside the DRAM buffer, whose access is contended by all the SSD channels, thus causing resources starvation.

6.1.2.3 Multi-plane RRAM

To reduce both the amount of commands processed by the SSD controller and the number of accesses to the internal DRAM, different RRAM page sizes were considered: 512 B, 1 kB, and 2 kB. To keep the payload of the 4 kB host transactions constant, SSD's firmware and RRAM memories were co-designed to work in multi-plane mode. In other words, when a n * 256 B RRAM page size is used, being $n = [2, 4, 8]$, the SSD's firmware is configured to read $16/n$ chunks of n * 256 B each from $16/n$ parallel channels. Figure 6.19 shows the cumulative latency distributions and the QoS of the simulated All-RRAM SSD as a function of the page size, when a host QD = 1 is selected. The optimal drive latency is achieved neither with the standard 256 B page size nor with the 4 kB NAND-like mode, but rather with a 1 kB multi-plane page configuration. Same considerations apply to other host QD values, as shown in Fig. 6.20a–d, there are two hot spots for the page size: when QD < 8 the value is 1 kB, whereas it becomes 2 kB for QD > 8. Figure 6.21 shows the

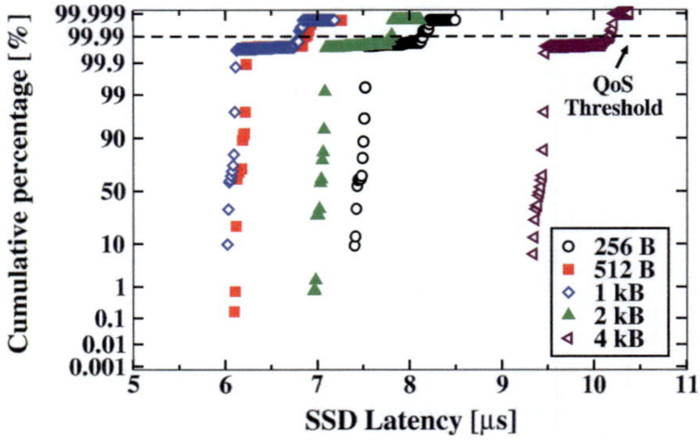

Fig. 6.19 Normal probability paper of the latency of the simulated All-RRAM SSD for QD = 1 and different page sizes

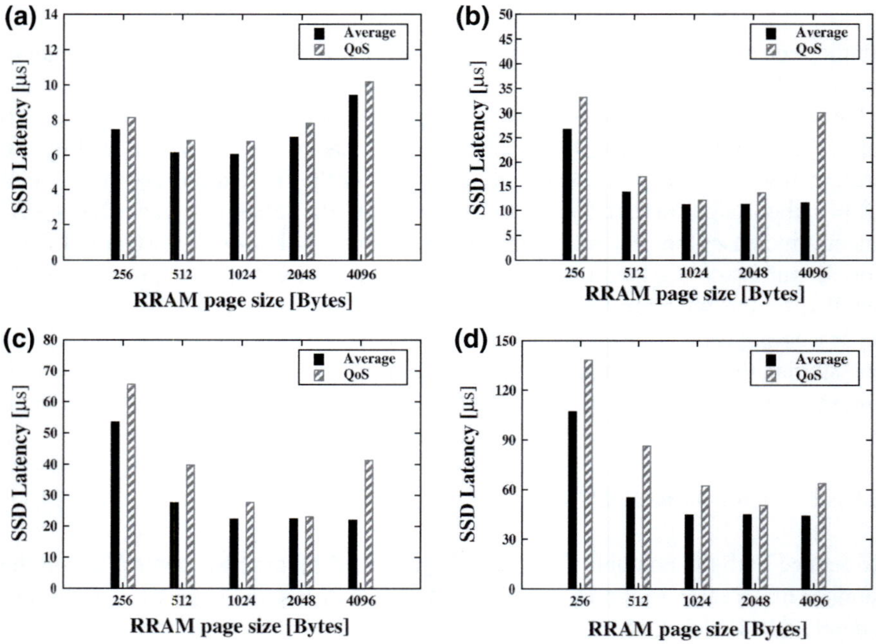

Fig. 6.20 SSD average latency and QoS for host QD = 1 (**a**), QD = 8 (**b**), QD = 16 (**c**), and QD = 32 (**d**)

bandwidth achieved by the All-RRAM SSD as a function of both the RRAM page size and the host QD. When QD = 1, the maximum bandwidth is with a page size of 1 kB; for QD > 16 the maximum bandwidth is with 4 kB.

These results proved that emerging memories such RRAMs have to be wisely designed when they are used as the main storage media in SSDs. In this regard,

Fig. 6.21 SSD's bandwidth as a function of the RRAM page size and the host QD

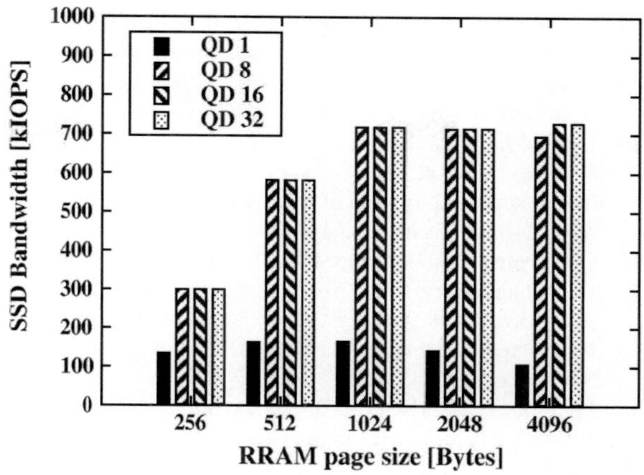

replacing NAND flash memories with RRAMs in a "plug and play" fashion is not the best way to reach high performances and low-latency. Moreover, even in the best working conditions (i.e., co-designing the memories characteristics together with the whole SSD architecture), it has been shown that the optimum design point of the All-RRAM SSD is still affected by the host configuration and its requirements. This is in agreement with today's trend of developing specific SSD architectures for specific host applications [16]; the downside of this approach is that it leads to extremely complex SSD designs with hundreds of parameters to explore. SSD-Explorer can definitely help to address the above mentioned problems, allowing a better understanding of where, in All-RRAM SSDs, the co-design activity is more effective.

References

1. R. Micheloni, A. Marelli, and K. Eshghi. *Inside Solid State Drives (SSDs)*. Springer, 2012.
2. L. Zuolo, C. Zambelli, R. Micheloni, D. Bertozzi, and P. Olivo. Analysis of reliability/performance trade-off in solid state drives. In *IEEE International Reliability Physics Symposium*, pages 4B.3.1–4B.3.5, June 2014.
3. E.I. Vatajelu, H. Aziza, and C. Zambelli. Nonvolatile memories: Present and future challenges. In *International Design Test Symposium (IDT)*, pages 61–66, Dec. 2014.
4. C. Zambelli, A. Grossi, D. Walczyk, T. Bertaud, B. Tillack, T. Schroeder, V. Stikanov, P. Olivo, and C. Walczyk. Statistical analysis of resistive switching characteristics in ReRAM test arrays. In *IEEE Int. Conf. on Microelectronics Test Structures (ICMTS)*, pages 27–31, Mar. 2014.
5. X. Y. Xue, W. X. Jian, J. G. Yang, F. J. Xiao, G. Chen, X. L. Xu, Y. F. Xie, Y. Y. Lin, R. Huang, Q. T. Zhou, and J. G. Wu. A 0.13 μm 8mb logic based cuxsiyo resistive memory with self-adaptive yield enhancement and operation power reduction. In *Symposium on VLSI Circuits (VLSIC)*, pages 42–43, June 2012.
6. K. Takeuchi. Hybrid solid-state storage system with storage class memory and nand flash memory for big-data application. In *IEEE International Symposium on Circuits and Systems (ISCAS)*, pages 1046–1049, Jun. 2014.
7. S.H. Jo. Recent progress in rram materials and devices. In *SEMICON Korea*, 2015.

8. S. Dubois. Crossbar Resistive RAM (RRAM): The Future Technology for Data Storage. In *SNIA Data Storage Innovation Conference*, Apr. 2014.
9. S Bates, M Asnaashari, and L. Zuolo. Modelling a High-Performance NVMe SSD constructed from ReRAM. In *Proc. of Flash Memory Summit*, Aug. 2015.
10. L. Zuolo, C. Zambelli, R. Micheloni, M. Indaco, S. Di Carlo, P. Prinetto, D. Bertozzi, and P. Olivo. Ssdexplorer: A virtual platform for performance/reliability-oriented fine-grained design space exploration of solid state drives. *IEEE Transactions on Computer-Aided Design of Integrated Circuits and Systems*, 34(10):1627–1638, 2015.
11. Open Nand Flash Interface (ONFI). http://www.onfi.org.
12. Ha Ryong (Harry) Yoon. Toggle-Mode NAND to Fill Growing Need for Higher Performance. In *Proc. of Flash Memory Summit*, Aug. 2009.
13. Nvm express 1.1 specification, 2013. http://nvmexpress.org/wp-content/uploads/2013/05/NVM_Express_1_1.pdf.
14. Intel solid-state drive dc s3700 series – quality of service., 2013. http://www.intel.com/content/www/us/en/solid-state-drives/ssd-dc-s3700-quality-service-tech-brief.html.
15. Open Nand Flash Interface (ONFI) revision 4.0. www.onfi.org/~/media/onfi/specs/onfi_4_0-gold.pdf?la=en.
16. Jian Ouyang, Shiding Lin, Song Jiang, Zhenyu Hou, Yong Wang, and Yuanzheng Wang. Sdf: Software-defined flash for web-scale internet storage systems. In *Conference on Architectural Support for Programming Languages and Operating Systems*, ASPLOS, pages 471–484, 2014.

Chapter 7
Simulation of SSD's Power Consumption

Lorenzo Zuolo, Cristian Zambelli, Luca Crippa, Rino Micheloni and Piero Olivo

Abstract In order to avoid changing existing infrastructures, SSDs need to fit in the same physical space of HDDs and, therefore, in their power envelope. Unfortunately, NAND power consumption is strongly dependent upon the Flash operation (i.e. read, write, erase). For many generations, SSD's power has been over- or under-estimated by adopting the worst case power consumption scenario (i.e. assuming write operations all the time); this approach is running out of steam because it always leads to exceeding power specs with the most recent NAND technologies. In order to correctly evaluate power, it is necessary to estimate the number of operations per time interval and their specific type (read, write, erase). Simulations are the only practical way to address this problem: when the prototype is ready (especially if the controller is a multi-million dollar ASIC) there are no chances to significantly reduce power, it's simply too late! This chapter shows how a SSD fine-grained simulator like SSDExplorer can be used as a design tool for developing power management algorithms that can minimize the power consumption of the entire SSD and, therefore, open the door to an even bigger adoption of SSDs (e.g. in data centers).

The adoption of SSDs in hyper-scaled environments such as cloud computing and big-data servers is opening a new era for storage, thanks to non-volatile memories. In this context, a major constraint that must be considered during the design phase of an SSD is its power consumption, which, as a matter of fact, limits the storage capacity of these devices. ExaBytes-dense (10^{18} Bytes) storage platforms completely relying on NAND flash memories (generally called "All-Flash" arrays) aim at completely replacing traditional HDDs [1–6]. However, to pursue this goal, besides the

L. Zuolo (✉) · C. Zambelli · P. Olivo
Dipartimento di Ingegneria, Università degli Studi di Ferrara,
via G. Saragat, 1, 44122 Ferrara, Italy
e-mail: lorenzo.zuolo@unife.it

L. Crippa
Microsemi, Via Torri Bianche 1, 20871 Vimercate, Italy

R. Micheloni
Performance Storage Business Unit, Microsemi Corporation, Vimercate, Italy

R. Micheloni (ed.), *Solid-State-Drives (SSDs) Modeling*,
Springer Series in Advanced Microelectronics 58,
DOI 10.1007/978-3-319-51735-3_7

standard chip power optimization of the SSD controller, it is mandatory to consider the entire storage infrastructure. When looking at SSDs, high capacity and power efficiency definitely don't go hand in hand, especially for general purpose workloads. For example, TLC NAND flash memories offer a large storage capacity but they cannot be used for write-intensive workloads because of their reduced endurance rate; moreover, they exhibit slower access time and higher power consumption compared to MLC and SLC. As a consequence, it is clear that applications have to be carefully co-designed together with the underlying storage system, by considering both the target performance and the power consumption simultaneously. This approach is leading to a new design methodology of SSD architectures called "software-defined Flash" [7]. This methodology tries to connect the application development with the SSD design phase, and to build a custom power-efficient and high-performing storage environment for a specific workload.

In this chapter a simulation methodology to assess the power consumption of a specific SSD architecture is presented. The main goal is to describe the different parts of the SSD design, and to highlight the increasing need to contextualize the design phase of the SSD inside its application scenario.

7.1 Accurate Estimate of the Actual SSD Power Consumption

As described in Chap. 1, NAND-Flash based SSDs feature an embedded controller and a parallel array of NAND Flash chips; if properly designed, this system allows achieving high throughput, low read/write latency and high reliability. However, to deal with the increasing performance requirements of hyperscale systems such as cloud computing and big data servers, SSD designers started to leverage the memory I/O parallelism, by adding hundreds of NAND memories to the same system. This approach leads to extremely fast yet complex solutions, which on one hand allow achieving the target performance/capacity of the drive but, on the other hand, heavily impact the overall SSD power consumption. This last point is of a particular concern in cloud environments due to the restricted power budget and to the severe green policies ruling their deployment [8].

To better understand where the power consumption issue comes from, several approaches have been proposed. Among them, it is worth pointing out 2 techniques. First of all, there is a top-down approach, which leverages the real hardware to measure the SSD's power consumption during its run time [9, 10]; on the opposite side, we have a simple bottom-up approach, which assesses the drive power consumption starting from that of NAND flash memories [11]. However, a closer analysis of these methodologies reveals that both present several drawbacks. From a designer perspective, measuring the power consumption of the drive after fabrication does not provide any significant value. Similarly, simulating the NAND flash power figures turns out to be inaccurate, especially considering that NAND program and read algorithms are not known and, hence, any power estimation could heavily differ from real silicon.

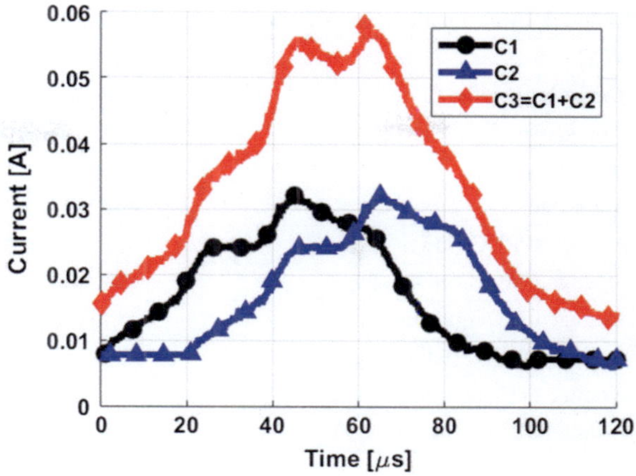

Fig. 7.1 Superposition principle of NAND currents. C1 and C2 are the currents measured on single chips, while C3 is the current measured on the main power supply

Assessing the power consumption during the design phase would be a desirable feature, which could help designers to efficiently define the drive architecture, thus avoiding any over-engineering. A possible alternative is a bottom-up approach that, starting from measured NAND flash currents during write and read operations, calculates the total SSD power consumption by using a cycle-accurate SSD simulator.

In this way, for any possible workload, the actual SSD power consumption can be estimated at any cycle. Furthermore, it is possible to estimate the percentage of power consumption related to flash memories and to other components (ECC, Flash controller, DRAM,...), as well as to foresee any possible overcoming of the maximum allowed power. The exact knowledge of the actual current sunk by a flash die at any single cycle allows taking actions to optimize the SSD power profile, such as, for instance, the program suspend.

The underlying assumption is that the Kirchhoff's current law holds true for the NAND flash sub-system. Since NAND flash chips are connected to the same power supply, their total current is, in first order approximation, equal to the sum of each NAND flash chip contribution. Figure 7.1 shows a current measurement done in a real SSD product which proves that this assumption is true. In this experiment, the SSD controller issued 2 read commands with a 20 μs skew to 2 TLC NAND flash chips, whose parameters are reported in Table 7.1, connected to the same power supply; C1 and C2 are the currents of the two chips, while C3 is the current of the main power supply. As anticipated, C3 is the sum of currents C1 and C2, and this proves the possibility of exploiting the superposition effect for estimating the NAND flash power consumption. To sum up, in order to simulate the power consumption of the whole NAND flash sub-system, these are the steps to follow:

- measure the current sunk by a single chip during standard read, program and erase operations;
- monitor the memories' ready/busy switching activity;
- combine the experimental current waveforms with the memory switching activity.

Table 7.1 Main characteristics of the measured NAND flash memories

VCC	3.3 V
Technology node	mid-1x-nm
Storage paradigm	TLC-128 Gb
Interface	Toggle
Bus speed	400 MT/s
Average tPROG (lower page)	800 μs
Average tPROG (middle page)	1900 μs
Average tPROG (upper page)	4300 μs
Average tREAD (lower page)	80 μs
Average tREAD (middle page)	100 μs
Average tREAD (upper page)	80 μs

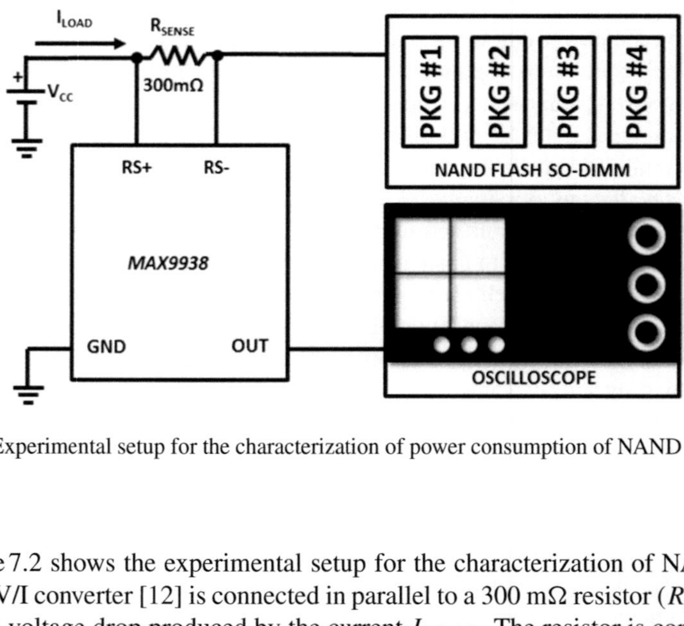

Fig. 7.2 Experimental setup for the characterization of power consumption of NAND flash memories

Figure 7.2 shows the experimental setup for the characterization of NAND currents. A V/I converter [12] is connected in parallel to a 300 mΩ resistor (R_{SENSE}) to sense the voltage drop produced by the current I_{LOAD}. The resistor is connected in series to the VCC power supply of the NAND-DIMM test board, where 4 different NAND flash chips are soldered (see Fig. 7.3). Both the sensing resistance and the current-Sense amplifier gain have been wisely selected to produce enough response bandwidth to capture all the current fluctuations during NAND Flash operations. Finally, the output current of the V/I converter is measured by an oscilloscope. The NAND-DIMM test board is connected to the custom developed *Automated Test Equipment* (ATE) shown in Fig. 7.4 through the interfacing socket. The ATE is composed by a programmable FPGA, a DRAM-buffer for data manipulation, two DIMM card sockets and two SAS ports for the connection to a computer and a power supply. The FPGA is programmed to behave as a simple NAND flash controller that

Fig. 7.3 NAND-DIMM test board

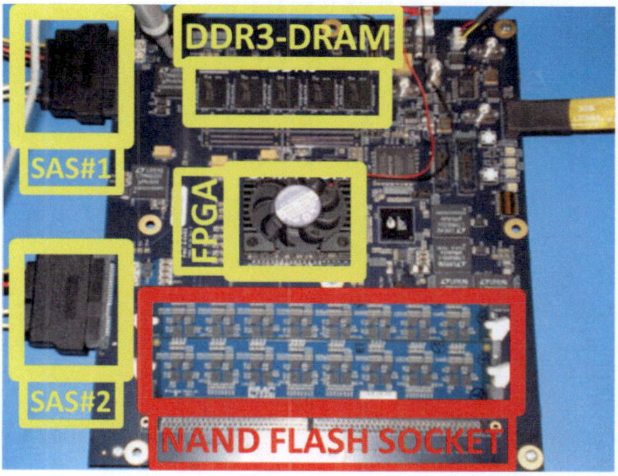

Fig. 7.4 ATE used for the experimental characterization of NAND flash memories

takes user commands as inputs and issues standard NAND flash read and program operations to all the chips.

Figure 7.5 shows the results of the power characterization of a mid-1x-nm TLC NAND flash. By looking at at the waveforms, it is possible to measure the timing of each operation and to identify all the current peaks. Especially for programming, the measurements allow capturing the current sunk during the Incremental Step Pulse Program algorithm (Chap. 2), which represents an extremely useful information when SSD power optimization algorithms have to be studied.

The NAND flash power measurement setup has been designed only for RBER characterization and power consumption assessment. In fact, the test equipment can only send raw IO/s to the memories under test and it is neither capable to execute a complete SSD firmware nor to handle the host interface protocol. Moreover, since it does not have any notion about the SSD architecture, it lacks of an intrinsic flexibility and, hence, it is not possible to explore any specific configuration without a new hardware implementation.

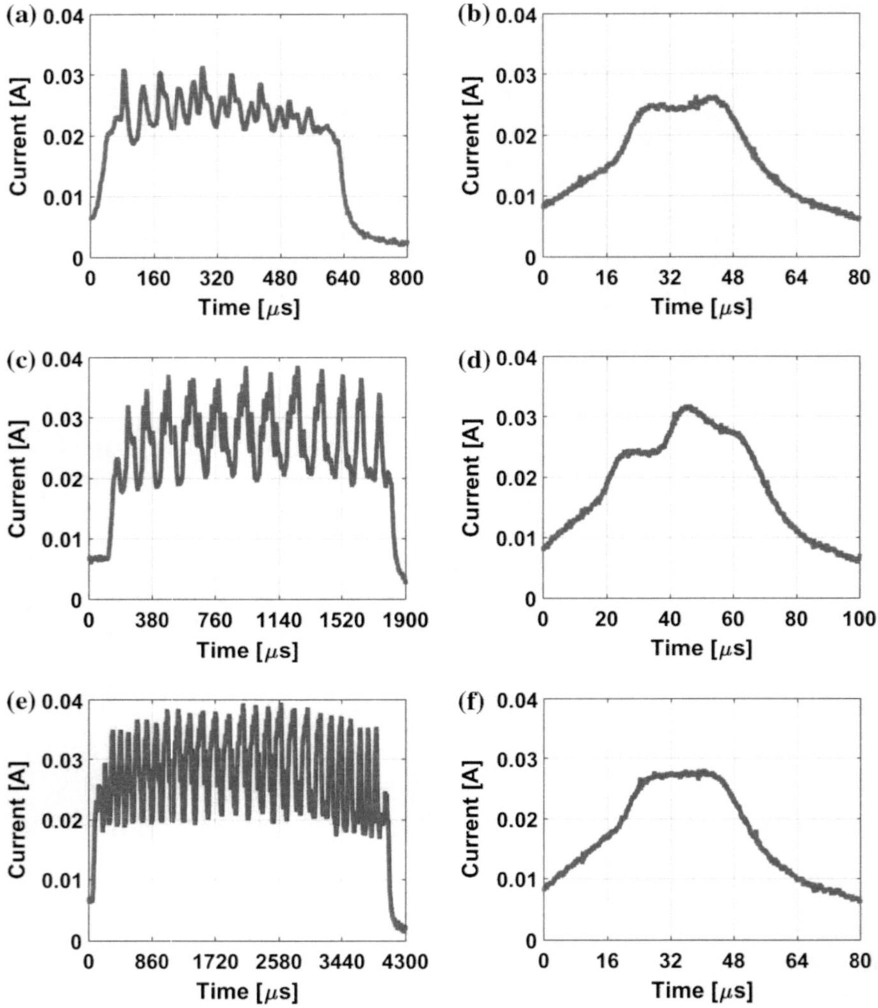

Fig. 7.5 Currents of a mid-1x-nm NAND flash chip measured during program and read operations. **a** lower page program; **b** lower page read; **c** middle page program; **d** middle page read; **e** upper page program; **f** upper page read

This problem has been addressed by exploiting a cycle-accurate SSD simulator, such as SSDExplorer. This tool, in fact, is able to monitor all the operations sent by the controller to the NAND flash memory array and to dynamically compute its ready/busy switching activity, as a function of the selected architecture and workload.

Figure 7.6 summarizes the main steps performed to estimate the SSD power consumption. Basically, the power traces are translated into a numeric format and sent to the power calculator engine. Given a specific drive architecture and host workload, the SSD simulator is in charge of estimating the switching activity of the Flash

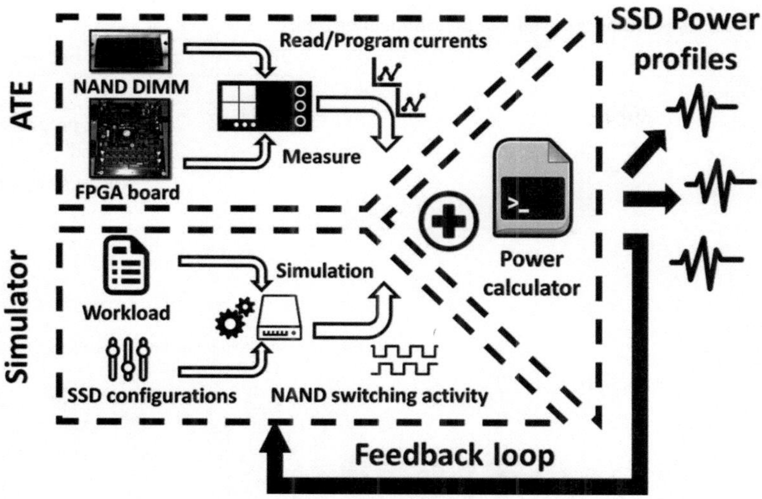

Fig. 7.6 Estimation of SSD's power based on real NAND flash current measurements and an SSD simulator

ready/busy. These contributions are then combined together by the power calculator engine which computes the final power profile, exploiting the superimposition principle. Indeed, it is possible to connect, in a single automated test flow, the design-space-exploration of latency and performance of a specific SSD architecture with the power consumption of its NAND flash memory sub-system. The described approach was applied to calculate the power consumption of the NAND flash sub-system; however, it can be applied to all the blocks of the SSD (Error Correction Code, core processors, DRAM, etc.), once the corresponding power traces are available.

One of the main advantages of this approach is that it can be used in a feedback loop with the SSD simulator to assess the power consumption of a target architecture and to dynamically modify the internal command sequence processed by the SSD. This iterative process can be used for optimizing the SSD power profile. For instance, uncontrolled multiple program operations issued on parallel chips may introduce out-of-spec current consumption, which represents one of the main problems that limits the internal parallelism and, therefore, the performance of an SSD architecture [13].

7.2 Optimization of SSD's Power Consumption

In order to prove this approach, we simulated the power consumption of an SSD with 16 channels, 8 NAND targets each [14], in the test configurations of Table 7.2. Figure 7.7 shows the power consumed by the NAND flash sub-system when test T1 is considered. In this case, average and peak power consumptions are 1.87 and 3.52 W, respectively, and the time required to process the workload is 660 ms.

Table 7.2 Tests performed to validate the simulation results

Workload (Random)	Test	Power optimization
25% Program 75% Read 512 Mbyte 4 kByte aligned	T1	None
	T2	Single program suspend
	T3	Double program suspend

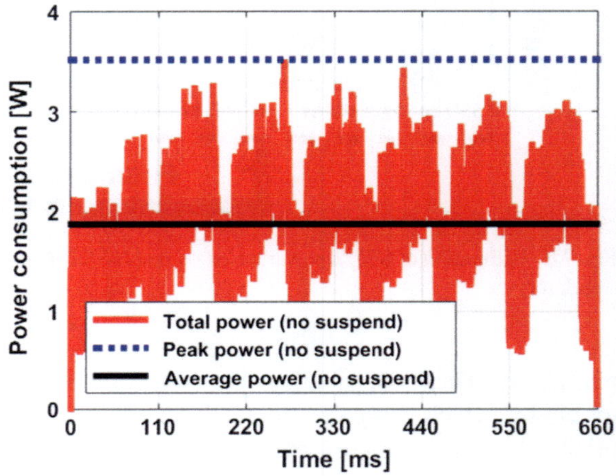

Fig. 7.7 Total power of the NAND flash sub-system during test T1 (used as baseline power consumption)

As it can be seen, even if random operations are issued by the host system, the power profile is somewhat repetitive. This is an expected behavior because random program operations are serialized by the SSD controller to follow the standard in-order programming sequence of NAND flash memories. After this test, which represents the reference configuration, the tool has been used to study a dedicated peak power management algorithm able to optimize the power profile of multiple program operations issued on multiple chips in parallel. The power consumption problem has been addressed through a power throttling algorithm, based on the program suspend operation [15].

In fact, modern Flash memories allow suspending a programming operation to perform a read. Once ready, the Flash controller can resume programming, without having to re-transfer the data to be programmed anymore (Chap. 2). Basically, depending on the power profile, the SSD controller can suspend any program operations to disoverlap current peaks from different chips. Moreover, to reduce the average power consumption of the drive, during program suspensions no other operations are issued to idle chips. Finally, considering that an upper page programming is the most power hungry operation (according to Fig. 7.5e), the peak power management algorithm has been applied to upper page programming only. During test T2 a single

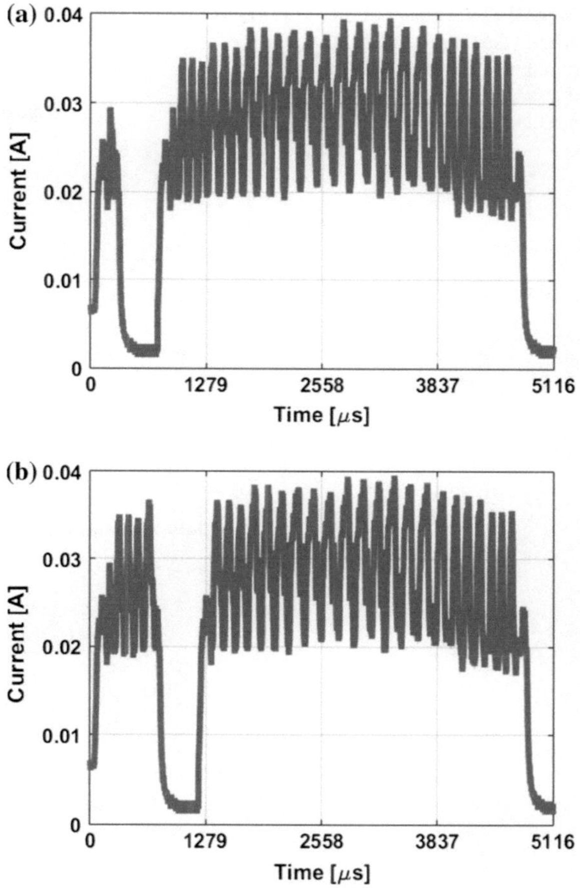

Fig. 7.8 Current profile during upper page programming when the timing of a single suspend window is optimized to reduce either **a** the average or **b** the peak current of the entire flash sub-system

suspend window of 200 µs is issued by the ATE during upper page programming. The resulting current profile is a function of when the Suspend command is issued.

In particular, we can "shape" the current profile of a single upper page programming operation to optimize either the average current or the peak current of the entire flash sub-system, as shown in Fig. 7.8a, b, respectively. A long series of experiments helped identifying the best conditions. Figure 7.9 shows the total power of the flash sub-system when either (a) the average current or (b) the peak current is optimized. With respect to test T1, the average and the peak power of the NAND flash sub-system were reduced of about 3.5 and 11.5%, respectively. As expected, however, the proposed approach negatively impacted the total processing time of commands, which increased of about 5%. This phenomenon is a direct consequence of the suspension window applied to upper pages, resulting in a longer latency of about 16%. During test T3 two suspend windows of 200 µs are issued by the ATE

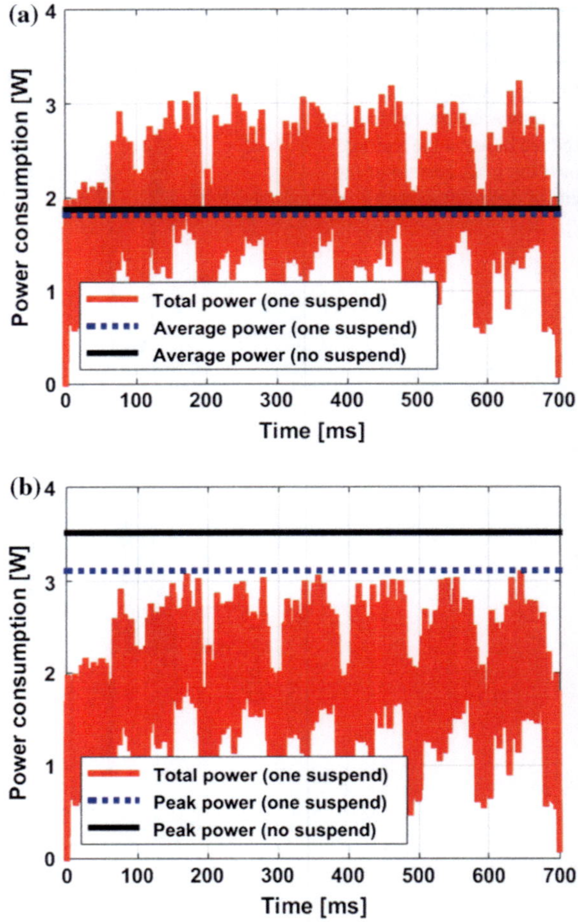

Fig. 7.9 Total power of the NAND flash sub-system during test T2 when either **a** the average current or **b** the peak current is optimized

during upper page programming. Also in this case, multiple experiments helped to find the best spot. The current profiles of a single programming operation that can optimize either the average current or the peak current of the whole flash sub-system are shown in Fig. 7.10a, b, respectively. Figure 7.11 shows the total power of the NAND sub-system in both cases. With respect to the baseline test T1, the average and the peak power of the NAND flash sub-system were reduced of about 5 and 13%, respectively. As expected, however, the proposed approach negatively impacted the total processing time of the drive which increased of about 10%; in fact, the latency of upper pages went up of about 27%.

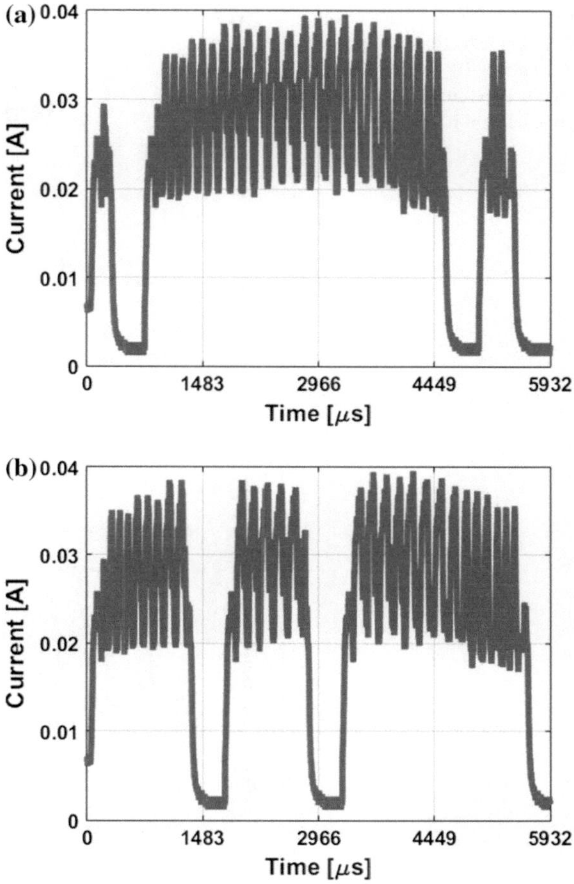

Fig. 7.10 Current profile during upper page programming when the timings of 2 suspend windows are optimized to reduce either **a** the average or **b** the peak current of the entire flash sub-system

The comparison between tests T1, T2 and T3 clearly highlights that power management algorithms trade a reduction of either the average or the peak power of the NAND flash memory sub-system with the total processing time of SSD commands.

SSDExplorer proved to be capable of managing power simulations thanks to its capability of monitoring how many memories are busy, and exactly when. As a matter fact, SSDexplorer can be used as a design tool for developing power management algorithms that can minimize the power consumption of the entire SSD and, therefore, it can open the door to an even bigger adoption of SSDs in data centers.

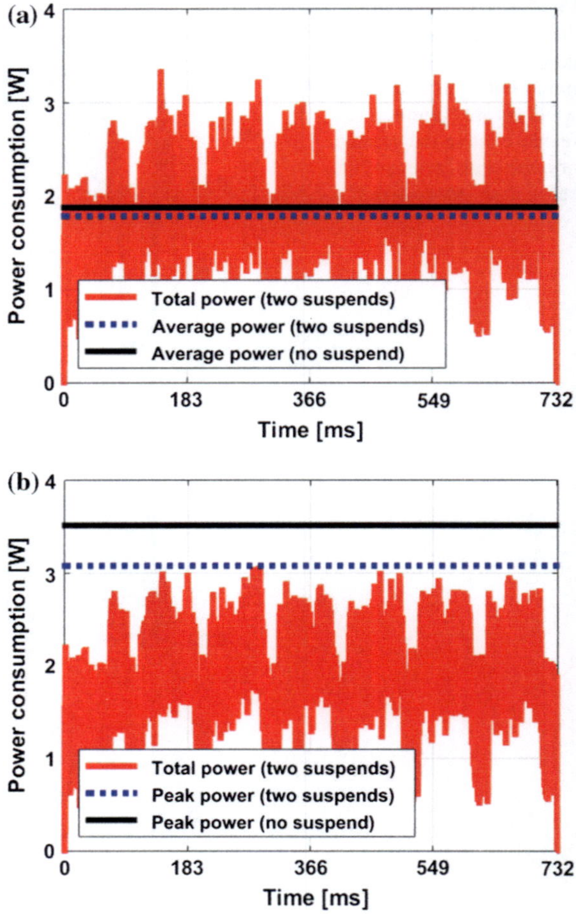

Fig. 7.11 Total power of the NAND flash sub-system during test T3 when either **a** the average current or **b** the peak current is optimized

References

1. Sean Barry. All Flash Array Data Protection Schemes. In *Proc. of Flash Memory Summit*, Aug. 2015.
2. Doug Rollins. Simplification: Get All Flash Performance Easily, Gradually, As Your Needs Grow. In *Proc. of Flash Memory Summit*, Aug. 2015.
3. Erik Ottem. All Flash Arrays in Healthcare. In *Proc. of Flash Memory Summit*, Aug. 2015.
4. Walter Amsler. All Flash Array Customer Case Study. In *Proc. of Flash Memory Summit*, Aug. 2015.
5. Avraham Meir. File on Flash: Delivering on the Promise of Webscale, All Flash, Distributed File Systems. In *Proc. of Flash Memory Summit*, Aug. 2015.
6. Somnath Roy. Ceph Optimization on All Flash Storage. In *Proc. of Flash Memory Summit*, Aug. 2015.
7. Jian Ouyang, Shiding Lin, Song Jiang, Zhenyu Hou, Yong Wang, and Yuanzheng Wang. Sdf: Software-defined flash for web-scale internet storage systems. In *Conference on Architectural Support for Programming Languages and Operating Systems*, ASPLOS, pages 471–484, 2014.

8. U.s. epa. report to congress on server and data energy efficiency. tech. rep., u.s. environmental protection agency, 2007.
9. Matias Bjørling, Philippe Bonnet, Luc Bouganim, and Bjorn Jonsson. uflip: Understanding the energy consumption of flash devices. *IEEE Data(base) Engineering Bulletin*, 33(4), 2010.
10. Balgeun Yoo, Youjip Won, Jongmoo Choi, Sungroh Yoon, Seokhei Cho, and Sooyong Kang. Ssd characterization: From energy consumption's perspective. In *Proceedings of the 3rd USENIX Conference on Hot Topics in Storage and File Systems*, pages 3–3, 2011.
11. V. Mohan, T. Bunker, L. Grupp, S. Gurumurthi, M. R. Stan, and S. Swanson. Modeling power consumption of nand flash memories using flashpower. *IEEE Transactions on Computer-Aided Design of Integrated Circuits and Systems*, 32(7):1031–1044, July 2013.
12. Maxim max 9938, 2011. http://www.maximintegrated.com/en/products/analog/amplifiers/MAX9938.html.
13. M. Sako, Y. Watanabe, T. Nakajima, J. Sato, K. Muraoka, M. Fujiu, F. Kouno, M. Nakagawa, M. Masuda, K. Kato, Y. Terada, Y. Shimizu, M. Honma, A. Imamoto, T. Araya, H. Konno, T. Okanaga, T. Fujimura, X. Wang, M. Muramoto, M. Kamoshida, M. Kohno, Y. Suzuki, T. Hashiguchi, T. Kobayashi, M. Yamaoka, and R. Yamashita. 7.1 a low-power 64gb mlc nand-flash memory in 15nm cmos technology. In *IEEE International Solid-State Circuits Conference - (ISSCC)*, pages 1–3, Feb 2015.
14. FLASHTEC NVRAM Drives. http://pmcs.com/products/storage/flashtec_nvram_drives/.
15. Jie Zhang, Mustafa Shihab, and Myoungsoo Jung. Power, energy and thermal considerations in ssd-based i/o acceleration. In *Proceedings of the 6th USENIX Conference on Hot Topics in Storage and File Systems*, HotStorage'14, pages 15–15, 2014.

Chapter 8
Simulations of the Software-Defined Flash

Lorenzo Zuolo, Cristian Zambelli, Rino Micheloni and Piero Olivo

Abstract In the last 30 years all software applications and *Operating Systems* (OSes) which make use of persistent storage architectures have been designed to work with HDDs. However, SSDs are physically and architecturally different from HDDs; in particular, SSDs need to execute a specific algorithm for translating host commands: the *Flash Translation Layer* (FTL). Basically, the main role of FTL it to mimic the behavior of a traditional HDD and to enable the usage of SSDs in any electronic systems without acting on the software stack. The main drawback of FTL is the *Write Amplification Factor* (WAF) which reduces both drive bandwidth and NAND flash reliability. Especially in the enterprise market and hyperscale data centers, performance and reliability losses induced by WAF are not tolerable. To deal with the above mentioned challenges, in the past few years software developers of hyperscale data centers have shown a growing interest for *Software-Defined Flash* (SDF), which leverages a new SSD design approach called *Host-based FTL* (HB-FTL). Thanks to SDF and HB-FTL it is possible to establish a native communication between the host system and the SSD. Basically, NAND flash memories can be directly addressed by the host without any further translation or manipulation layer. As a result, the write amplification phenomenon is removed and both performance and reliability are improved. However, according to the *Open-Channel* specifications, to enable the use of HB-FTL in SDF, the design of the SSD controller has to be slightly modified and its internal architecture has to be exposed to the host system. In this regard, the Open-Power initiative helps to provide a clear path to the HB-FTL execution, whereas the Open-Channel approach tries to define a standard for future SSD controllers, custom designed for SDF. This chapter describes a simulation

L. Zuolo (✉) · C. Zambelli · P. Olivo
Dipartimento di Ingegneria, Università degli Studi di Ferrara, via G. Saragat, 1,
44122 Ferrara, Italy
e-mail: lorenzo.zuolo@unife.it

R. Micheloni
Performance Storage Business Unit, Microsemi Corporation, Vimercate, Italy

R. Micheloni (ed.), *Solid-State-Drives (SSDs) Modeling*,
Springer Series in Advanced Microelectronics 58,
DOI 10.1007/978-3-319-51735-3_8

model that can be used to simulate SDF architectures exploiting the Open-Channel standard proposed by the Open-Power initiative.

In the last 30 years all software applications and *Operating Systems* (OSes) which make use of persistent storage architectures have been designed to work with HDDs [1]. However, SSDs are physically and architecturally different from HDDs; in particular, SSDs need to execute a specific algorithm for translating host commands: the *Flash Translation Layer* (FTL). Basically, the main role of FTL it to mimic the behavior of a traditional HDD and to enable the usage of SSDs in any electronic systems without acting on the software stack. Besides this translation operation, SSD controllers have to run garbage collection, command scheduling algorithms, data placement schemes, wear-leveling, and errors correction. All these routines, however, even if on the first hand allow a "plug and play" connection of the SSD with legacy hardware and software, on the other hand they do limit actual SSD performances. The main drawback of FTL is the *Write Amplification Factor* (WAF) which reduces both drive bandwidth and NAND flash reliability [2]. Of course, FTL and data manipulation are co-designed to reduce the WAF as much as possible; anyhow, a portion of the drive bandwidth has to be spent for FTL execution. Especially in the enterprise market and hyperscale data centers, performance and reliability losses induced by WAF are not tolerable.

To deal with the above mentioned challenges, in the past few years software developers of hyperscale data centers [3] have shown a growing interest for *Software-Defined Flash* (SDF). Indeed, in this kind of environments, the driving forces in the design of computational nodes are reliability and high performances: therefore, even the I/O management has to be re-architected. SDF leverages a new SSD design approach called *Host-based* FTL (HB-FTL) which allows the host system to:

- execute the FTL directly on top of its computational node;
- relieve the SSD from any host command translation or manipulation;
- reduce the WAF related to FTL execution.

Thanks to SDF and HB-FTL it is possible to establish a native communication between the host system and the SSD [4]. Basically, NAND flash memories can be directly addressed by the host without any further translation or manipulation layer. As a result, the write amplification phenomenon is removed and both performance and reliability are improved. However, according to the *Open-Channel* specifications [5], to enable the use of HB-FTL in SDF, the design of the SSD controller has to be slightly modified and its internal architecture has to be exposed to the host system. In this regard, the Open-Power initiative [6] helps to provide a clear path to the HB-FTL execution, whereas the Open-Channel approach tries to define a standard for future SSD controllers, custom designed for SDF.

This chapter describes a simulation model that can be used to simulate SDF architectures exploiting the Open-Channel standard proposed by the Open-Power initiative.

8.1 The Open-Channel Architecture

The Open-Channel architecture [7] aims at solving the WAF problem by introducing a lightweight system for SDF, where both user's applications and storage layer are NAND-flash aware [5]. Long story short, this new storage architecture enables HB-FTL by leveraging a more powerful processor located outside the SSD. This processing unit can be either the host processor itself or a dedicated accelerator in the form of a *Multi-Purpose Processing Array* (MPPA) [8, 9].

Figures 8.1 and 8.2 sketch the architectures that can be modeled by Open-Channel. Basically, thanks to the PCI-Express interconnection and the NVM-Express protocol [10], a bunch of NAND flash cards can establish a pear-to-pear communication with either the host processor or the MPPA, without requesting any specific management to the SSD controller [11]. In these architectures, "NAND flash cards" are not standard SSDs because, besides a simple I/O processor, a channel controller for NAND addressing and an ECC engine, they do not embody any complex processor, DRAM or even FTL (see Fig. 8.3). As a consequence, data read/write from/to these cards have to be considered as the raw output/input of NAND memories without any further manipulation.

One of the main advantages of this approach is that the host processor can view the MPPA either as a block device or as a programmable accelerator. In fact, using the I/O-pmem library [12] through the NVM-Express protocol, it is possible to address the DRAM modules hosted on top of the accelerator like a standard I/O block device. The MPPA, on the other hand, can manipulate and process the incoming data from the host, by exploiting specific applications hosted on top of its processors, thus enabling the "In-Storage Processing" approach [13]. As a matter of fact, the accelerator can execute these programs by means of a custom *Real Time Operating System* (RT-OS), designed to manage I/O operations more efficiently than the host OS. Moreover, along with the custom RT-OS, different FTLs can be executed by the MPPA, thus offloading either the non-volatile storage layer or the host processor from executing complex and latency/performance critical algorithms.

8.2 Simulation Model

A possible way to simulate the SDF based on the Open-Channel architecture is to use a programmable virtual manager able to virtualize all the peripherals required by the host system. In this regard, Qemu represents the preferable choice since it is open-source and it is ready to emulate complete OSes with several peripherals such as DRAM controller, Advanced Host Controller Interface (AHCI), Network Interface controller (NIC), and NVM-Express controller [14].

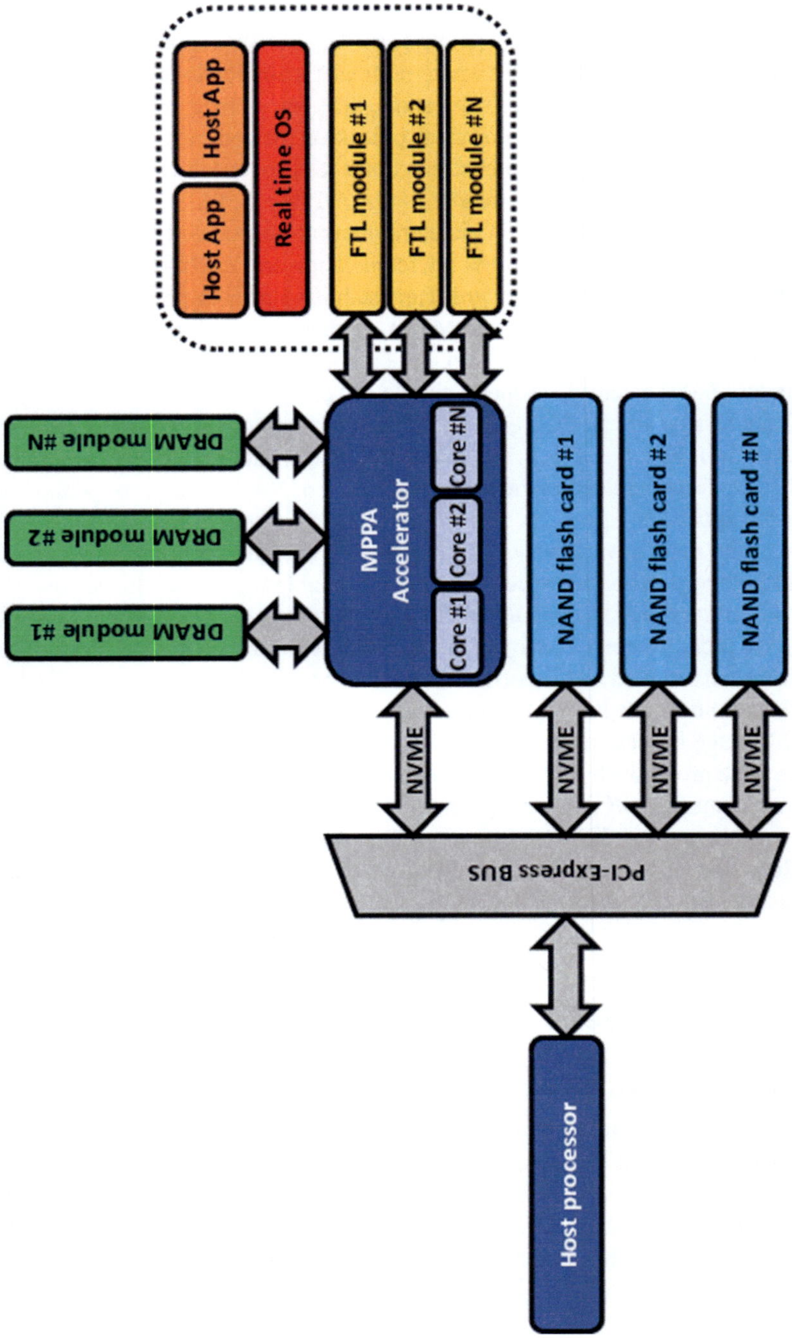

Fig. 8.1 Reference architecture modeled by the open-channel storage layer when a MPPA is used for HB-FTL execution

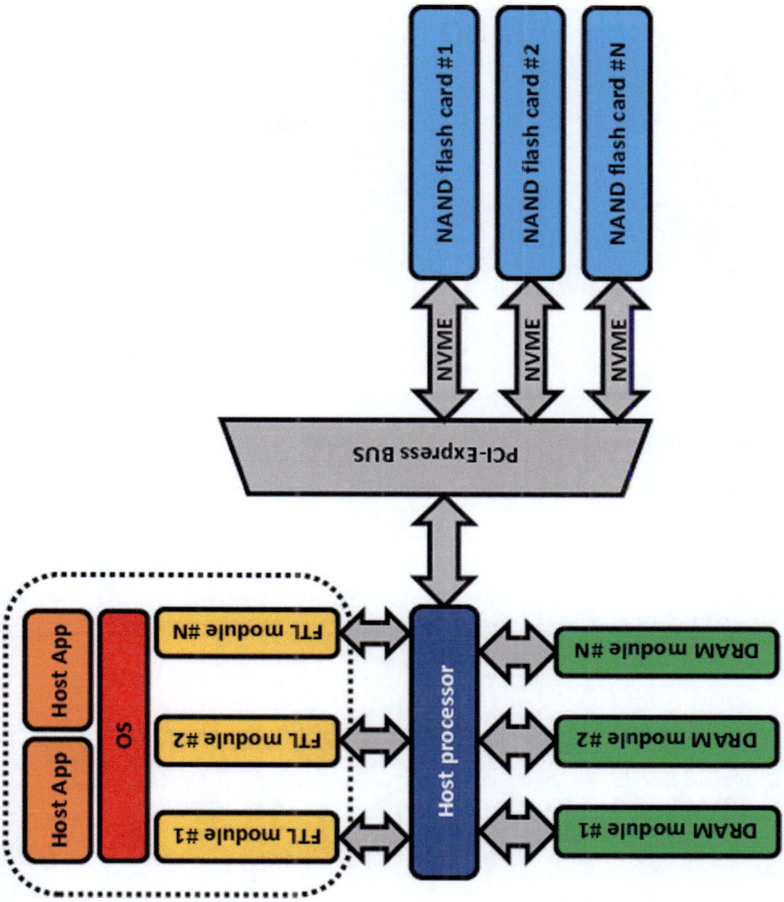

Fig. 8.2 Reference architecture modeled by the open-channel storage layer when the host processor is used for HB-FTL execution

As displayed in Fig. 8.4, the Qemu platform allows developing virtual peripherals which, once connected to the virtual device emulator layer, can be recognized as standard block devices. In this case, 2 general purpose block devices have been attached to the NVM-Express controller:

- the SSD block device: it acts as a storage I/O trace collector and can be used, for example, with the "Online-Offline" simulation mode presented in Chap. 3;
- the MPPA block device: it is a programmable, transactional-based functional simulator of the routines executed by the accelerator. Basically, the aim of this block is to provide a framework for "In-Storage Processing" and to introduce the proper delays for all the possible operations performed by the user.

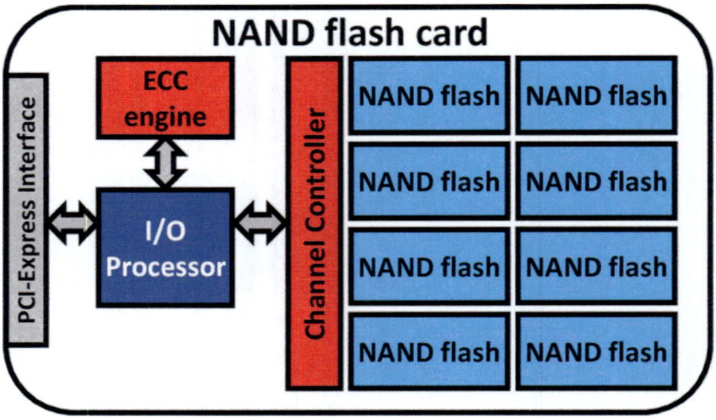

Fig. 8.3 NAND flash card used in the open-channel storage system

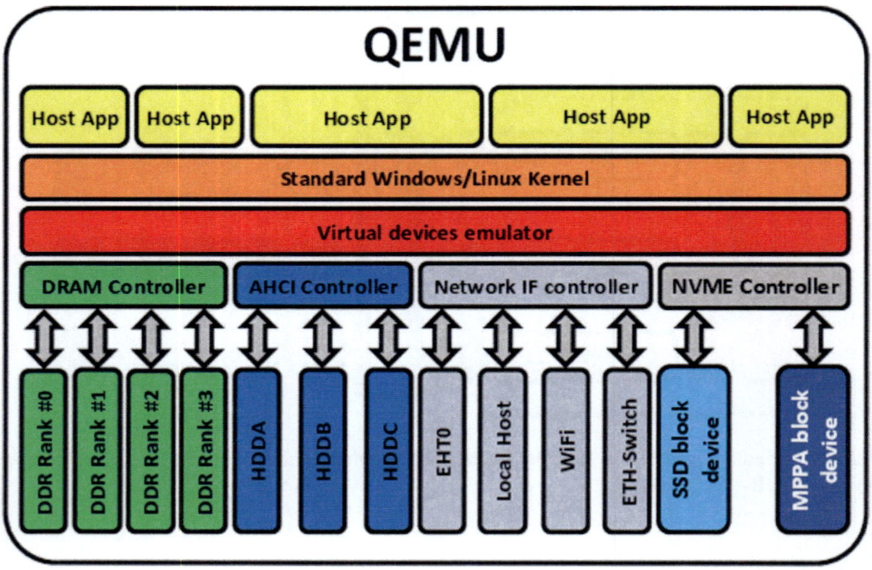

Fig. 8.4 Reference architecture modeled by Qemu. A custom I/O device for the SSD and the MPPA simulators are attached to the NVM-Express controller like traditional block devices

In other words, with a single integrated framework we can estimate how SDFs based on HB-FTL and Open-Channel impact the host system and user applications in terms of both performances and latency. Moreover, the use of Qemu for OS virtualization gives the possibility of developing and testing specific applications before real hardware prototyping, possibly suggesting new specs to both the Open-Channel architecture and the Open-Power initiative.

8.3 FTL Versus HB-FTL

In this section we provide a performance comparison between the Open-Channel architecture based on HB-FTL and a standard SSD design with embedded FTL. This comparison represents a crucial part in the validation process of the Open-Channel architecture: therefore, it is mandatory to collect data in the most realistic way. For this exercise, we decided to take a well known SSD architecture as a reference: the HGST SN150 Ultrastar [15]. Simulations have been performed by using SSDExplorer; following the specifications reported in Table 8.1, the whole drive architecture was simulated. Moreover, real workloads and storage I/O stack tracing have been provided by Qemu, which is connected to the SSD simulator itself.

Simulations cover 10 different mixed workloads, from a 100% 4 kByte random read—0% 4 kByte random write to a 0% 4 kByte random read—100% 4 kByte random write. In addition, to highlight possible subtle effects related to the steady state performance fluctuations introduced by a mixed read and write scenario, in each condition a 512 MByte workload has been simulated. These configurations represent the baseline for the following discussion; the FTL is executed by the SSD controller and neither the flash memories nor the SSD's architectural parameters are exposed to the host.

Figure 8.5 summarizes simulation results: in all the tested conditions the performance calculated by SSDExplorer is perfectly aligned with the one of the real device [16]. This result represents a good starting point for a fair evaluation of the Open-Channel and of the HB-FTL since no errors are introduced by the simulation framework.

As sketched in Fig. 8.6, to properly test SDF, which makes use of both the HB-FTL and the Open-Channel architecture, we used Qemu and added a software block miming both the FTL behavior and the SSD geometry (channels and targets) down to the low-level routines of a standard Linux OS image. In such a way, the OS can continue to use standard or custom file-systems, while address translation and data management required by the SSD are performed in background. Like for the FTL, the main purpose of the HB-FTL block is to translate the I/O performed by the host application in an SSD-compliant fashion. However, since this piece of software is no longer running within the SSD controller, but it is directly executed in the host system, additional tasks can be added. For example, in this case we introduced a data-

Table 8.1 HGST SN150 ultrastar configuration

Parameter	Configuration
Channels	16
Targets per channel	16
SSD capacity	3.2 TByte
NAND flash target	128 Gb Toshiba A19 eMLC
Host interface	PCI-Express Gen3 x4

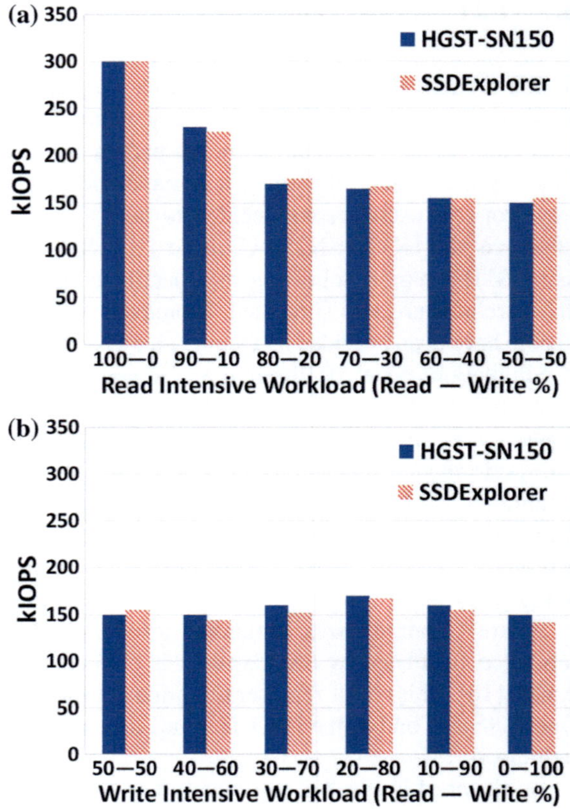

Fig. 8.5 HGST SN150 Ultrastar SSD: simulation accuracy of SSDExplorer for **a** read intensive and **b** write intensive workloads. A queue depth of 32 commands is used

merge algorithm which is responsible for keeping the host transactions aligned with the NAND flash block sizes of the SSD. Thanks to this block-write scheme, garbage collection caused by random write operations is completely removed (WAF = 1).

Figure 8.7 shows the simulation results of the HGST SN150 Ultrastar architecture. Compared to the conventional FTL, HB-FTL achieves highest bandwidth, especially with write intensive workloads. This is due to the fact that the Open-Channel architecture provides a straight path to flash memories; therefore, the OS can directly address the internal resources of the drive, removing, de-facto, the write amplification phenomenon of the garbage collection.

8.4 MPPA and HB-FTL

This section deals with the execution of the FTL on a dedicated MPPA card. This approach, hereafter denoted as HB-FTL-MPPA, makes use of an external hardware accelerator, connected to the PCI-Express bus, able to issue NVM-Express com-

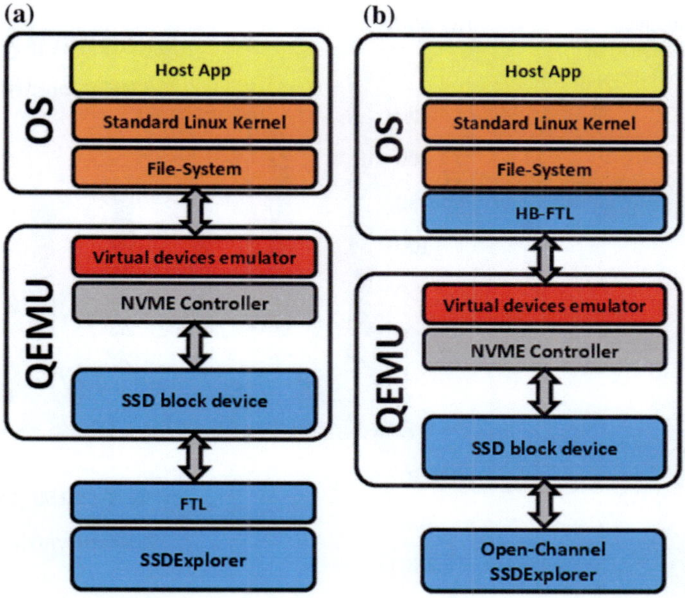

Fig. 8.6 Difference between legacy FTL and HB-FTL approaches

mands straight to the NAND flash cards. When the FTL runs on the MPPA (Fig. 8.1), all the considerations made in Sect. 8.3 about the HB-FTL implementation hold true; therefore, the write amplification phenomena induced by the garbage collection are completely absent. This time the main goal of the MPPA is the reduction of the I/O command submission/completion timings. These delays, which are strictly related to the host's processing capabilities, represent the time spent by the host to execute the NVM-Express driver and the OS file system for each submitted/completed I/O. Indeed, it has been demonstrated that the performance of NVM-Express SSDs are heavily affected by the I/O submission/completions timings [17]. Moreover, in most recent architectures like the one based on the 3D Xpoint technology [18], these delays can even represent the actual bottleneck of the whole storage layer, whose IOPS are limited by the host system itself. As a consequence, reducing these timings is key for designing ultra-high performance storage systems. A possible solution to this problem is to switch the NVM-Express protocol from an interrupt-driven I/O completion mechanism to a polling-driven approach. Basically, in standard NVM-Express-based SSDs, when an I/O is completed, the Flash controller sends an interrupt to the host notifying that the transaction is ready to be transferred/processed. After that, the host can submit another command to the drive because the submission of an I/O is driven by a completion event. In theory this approach requires the host take actions only when I/Os are submitted/completed, but in practice it introduces long processing timings because of the OS interrupt service routines [17]. Polling the I/O completion events can minimize the above mentioned processing timings:

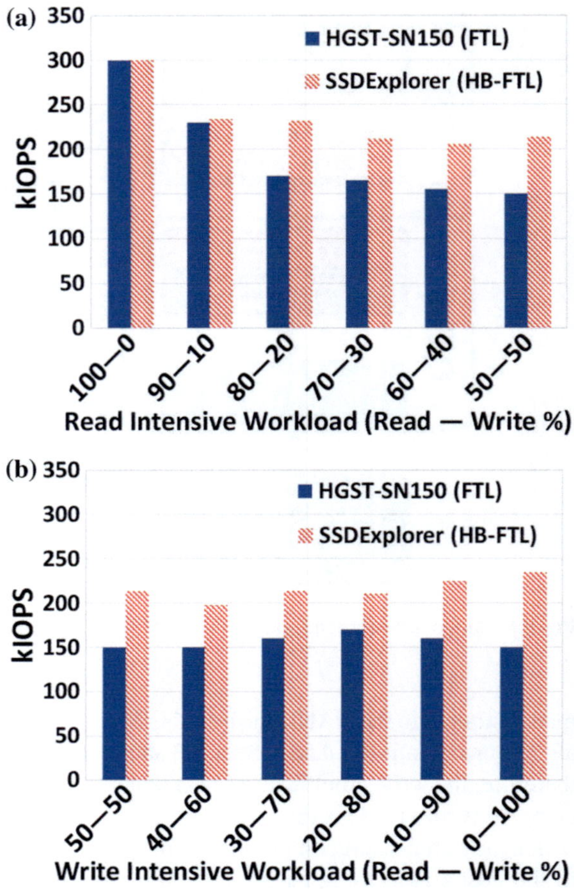

Fig. 8.7 Simulated throughput (kIOPS) of HGST SN150 Ultrastar SSD architecture when HB-FTL is used with **a** read intensive and **b** write intensive workloads. A queue depth of 32 commands is used

however, it requires that the host system continuously monitors the I/Os, thus wasting part of its processing capabilities. In light of all these considerations, moving the whole submission/completion process to a dedicated MPPA represents a good solution which can offload the host system and, at the same time, exploit the full performance of the NAND flash cards.

Figure 8.8 shows simulation results of the HGST SN150 Ultrastar for 3 different cases: off-the-shelf drive, drive with HB-FTL (Sect. 8.3), and drive with HB-FTL-MPPA. Results obtained with HGST SN150 Ultrastar in the off-the-shelf configuration are used as a basis for comparison.

To better understand the I/O acceleration capabilities of MPPA when used with Software Defined Flash, five different cases have been simulated, ranging from 95% speed-up of the host I/O submission/completion timings to 0% acceleration, i.e. no improvement. We decided to consider a maximum I/O acceleration of about 95% because of the hardware limitations introduced by the PCI-Express bus. The HB-FTL-MPPA is able to heavily improve performances in all the tested conditions,

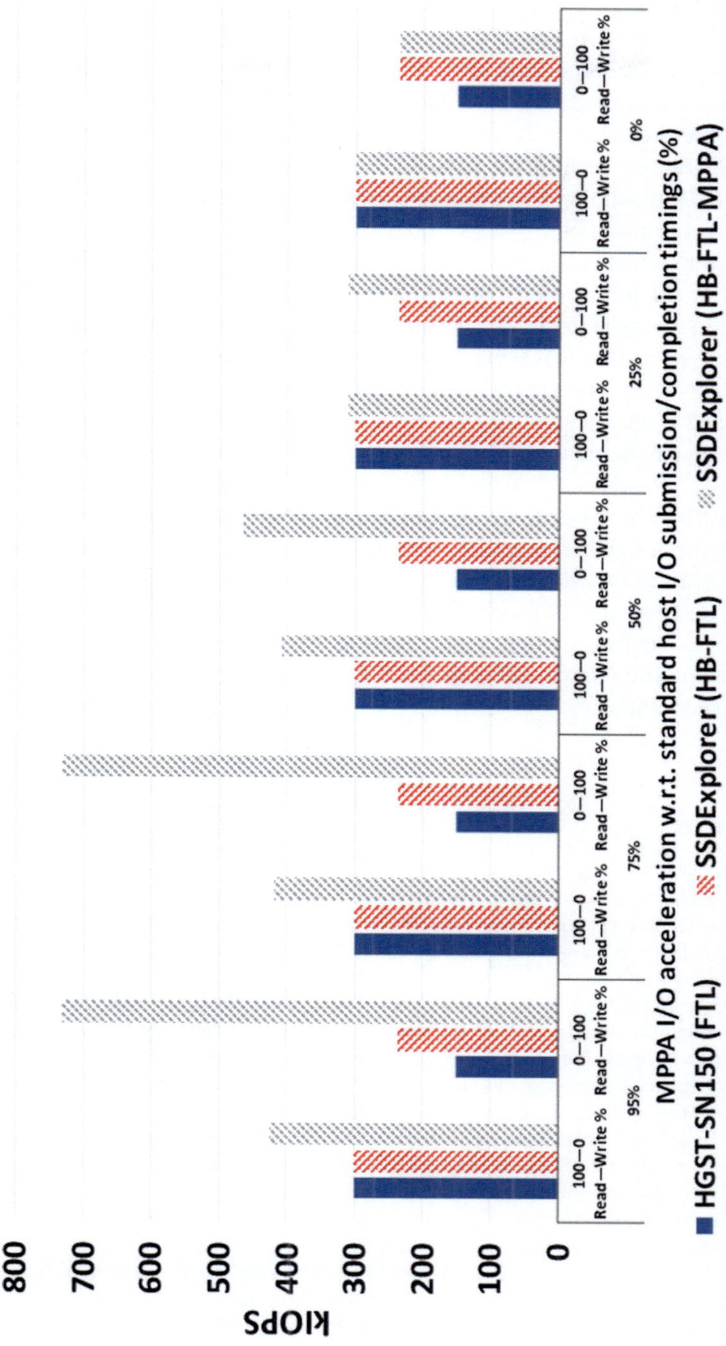

Fig. 8.8 Simulated throughput (kIOPS) of HGST SN150 Ultrastar SSD architecture when either HB-FTL or HB-FTL-MPPA are used for read intensive and write intensive workloads. Command queue depth is 32

but it is extremely effective when write intensive workloads are considered. This phenomenon is related to the fact that program operations on NAND flash cards still follow a *Write-Through* (WT) caching policy; therefore, once the data payload is transferred to the target card, a completion packet goes immediately back to the MPPA. At this point it is clear that, since the access time of WT buffers is in the order of a few microseconds, reducing the I/O submission/completion timings impacts the overall transfer time of the payload. This is also true for read operations, but because of the pipelining and queuing effects of the NAND flash cards, the overall improvement is not as big.

Simulation results clearly demonstrate that SDF, either as HB-FTL or as HB-FTL-MPPA, can improve the bandwidth and reduce the write amplification effects compared to a conventional SSD design. It is worth highlighting that these achievements represent just the first step towards a new SSD design methodology: a complete virtualization of the storage backbone might become real pretty soon. In fact, both HB-FTL and Open-Channel allow to virtually separate the internal resources of the SSD (like channels and targets), providing a clear and straight path to OS data partitioning.

Based on the above mentioned results, SSDExplorer proved to be a great candidate for fueling the design of future SDF. Indeed, thanks its easy customization, both SSD designers and software developers can efficiently co-design the internal resources of the drive, tailoring them to the actual application requirements. Last but not least, SSDExplorer can link the user applications to the actual power consumption of the NAND flash cards, allowing to optimize not only the performances of the system by also its power efficiency.

References

1. R. Micheloni, A. Marelli, and K. Eshghi. *Inside Solid State Drives (SSDs)*. Springer, 2012.
2. Xiao-Yu Hu, Evangelos Eleftheriou, Robert Haas, Ilias Iliadis, and Roman Pletka. Write amplification analysis in flash-based solid state drives. In *Proceedings of SYSTOR*, pages 10:1–10:9, 2009.
3. Jian Ouyang, Shiding Lin, Song Jiang, Zhenyu Hou, Yong Wang, and Yuanzheng Wang. Sdf: Software-defined flash for web-scale internet storage systems. In *Conference on Architectural Support for Programming Languages and Operating Systems*, ASPLOS, pages 471–484, 2014.
4. Ashish Batwara. Leveraging host based Flash Translation Layer for Application Acceleration. In *Flash Memory Summit*, 2012.
5. Javier González, Matias Bjørling, Seongno Lee, Charlie Dong, and Yiren Ronnie Huang. Application-driven flash translation layers on open-channel ssds. In *Proceedings of the 7th Non Volatile Memory Workshop (NVMW)*, pages 1–2, 2016.
6. Open Power Ready Definition and Criteria. http://openpowerfoundation.org/wp-content/uploads/resources/openpower-ready/content/ch_preface.html.
7. Open-channel solid state drives, 2016. http://openchannelssd.readthedocs.org/en/latest/.
8. The kalray multi-purpose-processing-array (mppa), 2016. http://www.kalrayinc.com.
9. Patrice Couvert. High Speed IO processor for NVMe over Fabric (NVMeoF). In *Flash Memory Summit*, 2016.
10. NVM Express, 2013. http://www.nvmexpress.org/.

11. S Bates. Accelerating Data Centers Using NVMe and CUDA. In *Proc. of Flash Memory Summit*, Aug. 2014.
12. Persistent memory programming, 2016. http://pmem.io.
13. Yang Seok Ki. In-Storage Compute: an Ultimate Solution for Accelerating I/O-intensive Applications. In *Flash Memory Summit*, 2015.
14. QEMU: open source processor emulator. http://wiki.qemu.org/Main_Page.
15. Ultrastar SN150 Series NVMe PCIe x4 Lane Half-Height Half-Length CardSolid-State Drive Product Manual. https://www.hgst.com/sites/default/files/resources/US_SN150_ProdManual.pdf.
16. HGST Ultrastar NVMe SN150 Enterprise SSD Review. http://www.tomsitpro.com/articles/hgst-ultrastar-nvme-sn150-enterprise-ssd,2-996.html.
17. Yang Jisoo, B. Minturn Dave, and Hady Frank. When polling is better than interrupt. In *USENIX Conference on File and Storage Technologies, (FAST'12)*, 2012.
18. Frank Hady. Wicked Fast Storage and Beyond. In *Proceedings of the 7th Non Volatile Memory Workshop (NVMW)*, 2016.

Index

R. Micheloni (ed.), *Solid-State-Drives (SSDs) Modeling*,
Springer Series in Advanced Microelectronics 58,
DOI 10.1007/978-3-319-51735-3

MIX
Papier aus verantwortungsvollen Quellen
Paper from responsible sources
FSC® C105338

If you have any concerns about our products,
you can contact us on
ProductSafety@springernature.com

In case Publisher is established outside the EU,
the EU authorized representative is:
Springer Nature Customer Service Center GmbH
Europaplatz 3, 69115 Heidelberg, Germany

Printed by Libri Plureos GmbH
in Hamburg, Germany